D1141057

ONE SMALL STEP

ONE SMALL STEP

THE APOLLO MISSIONS

THE ASTRONAUTS

THE AFTERMATH

A 20 YEAR PERSPECTIVE

TIM FURNISS

Dedication:
To my wife Sue, with love
at T + 10 years

A **Foulis** Space Book

First published 1989

Published by:
Haynes Publishing Group
Sparkford, Nr. Yeovil, Somerset
BA22 7JJ, England.

Haynes Publications Inc.
861 Lawrence Drive, Newbury Park,
California 91320, USA.

British Library Cataloguing in Publication data
Furniss, Tim,
One small step.
1. American astronauts
I. Title
629.45'0092'2

ISBN 0-85429-586-0

Library of Congress catalog card number 89-83552

Editor: Mansur Darlington
Design and Layout: Mike King
Printed in England by: J. H. Haynes & Co. Ltd

Contents

Acknowledgements

ALL PHOTOS ARE FROM NASA UNLESS indicated otherwise in the captions.

I would like to thank David Shayler, a space author and proprietor of Astro Information Service, High Farm Road, Halesowen, West Midlands, B62 9RX, England, and manned spaceflight researcher and historian Keith T. Wilson for their help in providing information and data. Thanks also to NASA historian Lee Saegasser in Washington DC for his assistance and to Lisa Vazquez of NASA, Houston and Douglas Arnold for finding some elusive photos.

All twelve moonwalkers were sent letters with requests for latest biographical information and photos plus a blank tape and a set of questions. Only one, Jim Irwin, responded fully and for that I am most grateful. Neil Armstrong's personal assistant provided his latest biographical information sheet as did Pete Conrad's. Alan Shepard sent his latest biographical information and a photograph. Ed Mitchell responded helpfully with information. John Young responded through a NASA public affairs officer that he was not in a position to assist. Charlie Duke was very helpful and responsive. Gene Cernan did not feel able to assist with taped responses to questions sent by post but was amenable to a personal interview if that were possible, which in the end it was not. Jack Schmitt declined my request for assistance by way of asking for a $4000 a day fee. I failed to get a response from Buzz Aldrin, Alan Bean or Dave Scott and I feel that my inability to make contact was more a reflection of being in the wrong orbit, rather than any unhelpfulness on their part, although, ironically, I met Buzz Aldrin after I had finished this book!

Research was made using NASA press kits, mission reports, voice transcripts, US military services biographies, contemporary media coverage and personal observation.

There are a great many publishers to whom thanks are due for their kind permission for me to refer to previously published material and in the case of newspapers, to reproduce headlines, in order to recreate the mood of the 'Moon Race' and the unique character of each Apollo moon mission. A list of quoted material and sources can be found at the back of this book.

On completing the manuscript I sent each 'moonwalker' a copy of his own biographical piece for comment. Alan Shepard, John Young and Jack Schmitt declined even to acknowledge receipt of the material, even after being requested to do so twice – sadly, their comments could not, therefore, be included.

The remainder responded most helpfully in correcting inaccuracies or misunderstandings, or in providing new information. To them, my thanks.

Genesis

ON SUNDAY EVENING, 20 JULY 1969, LIKE millions of other people in other times zones across the world, I was watching TV. Unlike any other day before, we were all watching coverage of the same event, a rarer occasion than it is today, as it was only four years since the launch of Early Bird, the comsat that fuelled the communications revolution from geostationary orbit which began with Syncom a little earlier. For someone who had followed the space race of the sixties with a boyish enthusiasm – some said fanaticism – to match my tender years, the first manned landing on the Moon had been the greatest event in my life. The whole world, it seemed, was captivated by what President Nixon described at the time as the "greatest week since creation" – the whole world that is, except historian A. J. P. Taylor. During ITV's coverage of the first landing, as the world was waiting for the astronauts to make the first steps onto the surface, he was asked by anchorman David Frost what he thought the historical significance of it would be. Mr Taylor replied that it would be seen as the most irrelevant, extravagant event in history. His remarks were greeted by howls of derision from the studio audience and no doubt by most of those who weren't watching the other side, the BBC, and the ebullient James Burke – and his infernal electronic pointer.

Only twelve men walked on the Moon, between July 1969 and December 1972. The last unmanned spacecraft to land there arrived in 1976. Nothing has returned since. To the younger generation, who were not born or were to young to remember Apollo, the project must seem like it did to Mr Taylor. In retrospect, it seems like a goldrush to nowhere, the gold being the $25 billion spent to get there. Maybe Mr Taylor was right.

One can identify many events in history that were the genesis of man's exploration of the Moon, from the sense of wonder of the caveman to the stories of Jules Verne. Man was always going to get there – eventually. And when he did a team of scientists would establish a large moon base for lengthy exploration of the surface. Instead, Man rushed there as quickly as possible, stayed for the minimum period and came home, never to return. Instead of an initial, explorative foray, followed by more detailed and significant excursions, Apollo was IT. So, the real genesis of man's exploration of the Moon is, in fact, only the genesis of the Apollo programme. As things turned out this was not an undertaking spurred by man's romantic vision and sense of wonder but by political panic riding on an intercontinental ballistic missile.

Launched on 3 August 1957, it was the first – and it was also *Russian*. The event sent shivers up American spines. The Soviets had

Project Apollo was conceived as a result of the American reaction to the launch of this: Sputnik 1, on 4 October 1957. (Novosti)

the capability to deliver a nuclear weapon onto America's doorstep in a matter of minutes. America had no reply: its Atlas ICBM was still languishing on the launch pad. And the Atlas was less powerful too, as America was to learn to its chagrin in October 1957.

Two years earlier, the United States announced that it would launch a small science satellite on a non-military rocket called Vanguard. The satellite would be the USA's participant in the International Geophysical Year, which was to take place between July 1957 and December 1958. The satellite was to be the star among mere high altitude balloons, sounding rockets and aircraft. But also in 1955, the Soviet Union announced that it would launch a satellite. This would weigh 184 pounds – against Vanguard's three – the Soviets said.

No one, it seems, took any notice. Internal political problems in the US space team stole the limelight. German rocket engineer Wernher von Braun pleaded with President Eisenhower to let him launch the first satellite, which would be heavier than Vanguard, on a version of his intermediate range ballistic missile called Redstone. Eisenhower would have none of it. A military missile launching a science satellite? Four days after the unannounced launch of the Soviet ICBM, von Braun fired his Redstone, now with upper stages and renamed Jupiter C. The missile reached an altitude of 600 miles and could have placed the world's first satellite in space had Eisenhower not forbidden the exasperated German to equip the Jupiter with a final stage. Meanwhile Vanguard was having serious problems and von Braun's blood was boiling.

As advertised, the Soviet Union duly launched its heavyweight satellite, called Sputnik, on 4 October 1957. Von Braun was dining with associates at Huntsville in Alabama, when someone burst into the room to break the news. The memory of the intensity of his pained expression never left those who witnessed it. The launch of Sputnik 1 sent shockwaves around the world but it needn't have. Americans demanded to know where *their* ICBM was, and when *their* satellite was going to be launched. More significantly, the military were pained at the realisation of the power of the Soviet's ICBM, which could place 184 pounds into orbit. Ironically, that 184 pound satellite was a replacement for the original one which weighed 1,500 pounds but was not ready in time. The US policy of delivering nuclear weapons by air using Hustler bombers had backfired miserably and the development of the Atlas ICBM, which had been rather on a backburner, was injected with afterburner. Von Braun again pleaded with Eisenhower to let him restore America's lost pride with the fully equipped Jupiter C. Vanguard would make the first attempt, said Eisenhower from the golf course. He just hadn't taken Sputnik seriously, nor its implication.

Worried Vanguard project officials hadn't even test fired the three stage version of the satellite launcher when, on 3 November 1957, the Soviets launched Sputnik 2. This time, the satellite remained attached to the final stage of the rocket and, as a result, over 1,000 pounds was orbiting the Earth. The craft was large enough to radiate sufficient reflected sunlight for it to be seen in the night sky. Millions peered in awe at the wandering star, from the native in Africa who had no idea what it was, to the Devon farmer who enthusiastically hauled my family and me from his farmhouse in April 1958, shouting in unmistakable west of England tones, "there she goes!" "She", in fact, was the the canine bitch Laika who mercifully had been longtime dead, as Sputnik was in its death throes before an imminent re-entry. While Laika, whose air ran out six days after launch and clearly was not coming home, captured the attention of animal lovers all over the world, that a living being was in orbit and was Russian to boot was a far more significant fact to the US space team. Even Eisenhower seemed to be taking notice from the nineteenth tee, keen to see his country's reply.

In early October 1957, Vanguard's project officer, John Hagen, briefed the president at the White House. He said that the first test firing of the Vanguard rocket with all three live stages would take place in early December and that if the third stage worked perfectly then there was just the chance, repeat chance, that a small instrumented satellite would be placed in orbit. Hagan returned to his office in the belief that the briefing had been confidential. On the launch pad at Cape Canaveral – a mosquito infested sand spit jutting out of the east coast of Florida – Vanguard was being prepared for its first full test trial. To his horror, Hagan heard on the radio a White House spokesman announce that the first US satellite would be launched on 3 December. He could have done without that sort of pressure.

Third December arrived and the press flocked to Cocoa Beach, assembling their cameras to point at the southern part of the sand spit only a mile or two across the water. Americans watched on TV. The world listened on the radio. The slender, rather elegant, rocket ignited and rose, to the roar of its engine and to the encouragement of observers. Three seconds into its epic mission, Vanguard returned perfectly to the launch pad and broke apart amid a cataclysmic explosion. Comically, its tiny satellite fell away from the conflagration and was found later bleeping away merrily, totally unscathed. This had to be the lowest moment for America. True to form, the nation, in that rather naive way it sometimes behaves, placed all its dirty space washing on the line for all to see, as it anguished over the disaster. Thanks to some inspired newspaper sub-editors, the event became known as 'Kaputnik' and 'Flopnik'. One US politician who pounded his fist at Eisenhower and demanded some positive action in the form of a co-ordination sense of direction, was senator Lyndon Baines Johnson, whose influence was to become a critical factor in America's future in space.

During 1958, the Soviet Union launched just one satellite, a laboratory called Sputnik 3. It also suffered some launch failures but because its programme was a secret one, nobody in the west knew. On the other hand, the USA finally got a satellite, Explorer 1, into orbit, thanks ironically to the von Braun team and a Jupiter C, and went on to place four other satellites into orbit. Two spacecraft were successfully placed *en route* for the Moon, one of them rather optimistically aiming to become a lunar orbiter, but both fell back to Earth from thousands of miles up because their carrier rockets failed to attain enough velocity.

But also in 1958, very publicly, the US had suffered eight failures, the most spectacular of which occurred when the first lunar probe attempt ended in an explosion 77 secs after the launch of its Thor Able booster. It was these failures that people remembered, not the successes of Explorer 1 in discovering the Van Allen radiation belts, Vanguard – finally space borne – testing the first solar cells for electrical power generation, Pioneer reaching deepest space, or Score becoming the first 'communications' satellite, by broadcasting repeating messages

America's hopeful seven astronauts. Who would be first? From left to right: Grissom, Shepard, Carpenter, Schirra, Slayton, Glenn and Cooper.

of Christmas cheer from Eisenhower to anyone able to receive them anywhere in the world, Despite its one success with Spunik 3, the Soviet Union, it seemed, was still in the driving seat.

This apparent trend continued in 1959. Despite secret launch failures, the Soviets flew three Lunik spacecraft towards the Moon. The first failed to hit its target but became the first artificial planet, much to the admiration of the world. Lunik 2 became the first man-made object to hit the Moon, and the spectacular Lunik 3 flew a figure eight path around the Moon, photographing its far side, revealing for the first time in history what the human eye had not seen before. The USA launched eleven satellites and space probes, including the first US solar orbiter, but suffered eight public launch failures, five of them consecutively between April and July. The most bizarre and spectacular of all, occurred when a Juno 2, carrying Explorer S1, turned tail above the launch pad and had to be destroyed. During that year, the USA had flown its first mission of the X-15 rocket plane, a precursor to a reusable manned spaceplane which could be flying within five to ten years.

It was becoming clear by now that space had become a theatre. The feats and failures of space had captured the imagination of the world and the competition between the Soviets and the USA had become the Space Race, the implication being that he who lead in space, lead militarily and technologically on Earth, such an important factor during the Cold War which existed at the time. The pace was quickening. Whilst the plan to launch spacemen in reusable, versatile spaceplanes may have been seemed logical – and, indeed wholly practicable – it wouldn't make a nation the first to place a man in space. The emphasis now was to launch a man in space as quickly as possible. At one time, indeed, the USA even called the project Man in Space Soonest, or MISS. Thus, instead of using the spaceplane, the first spacemen would be launched in the sophisticated nosecones of brutal intercontinental ballistic missiles or the slightly more gentle intermediate range ballistic missiles. A new space agency, called NASA, appointed seven military test pilots who would fly the Mercury capsule. These seven became legends in their lifetimes, their training coming under the most public of gazes. But before they flew, chimpanzee test pilots would clear the way on a number of pathfinding missions. Over in Russia, twelve cosmonauts were selected secretly and their path to space paved by dogs.

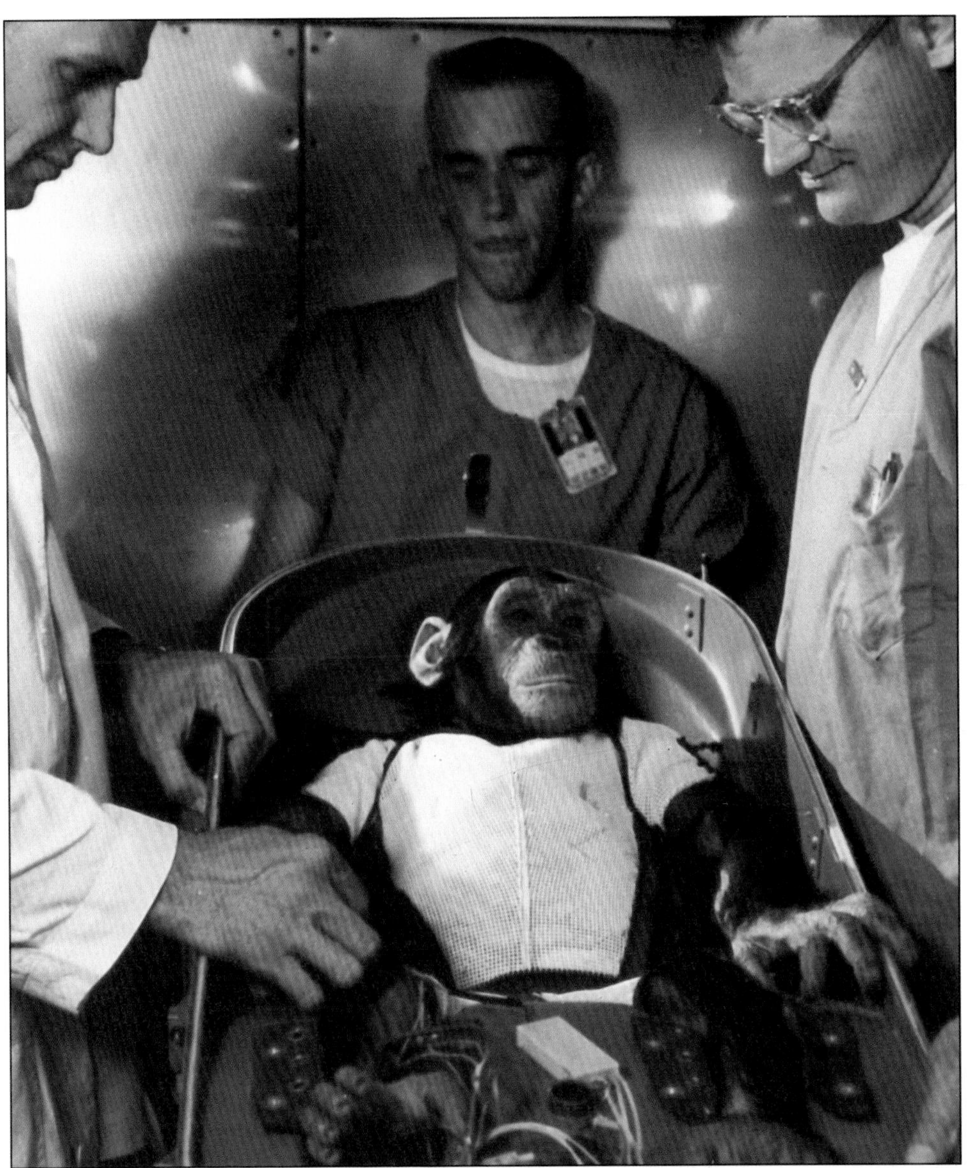

Pioneer astronaut Ham.

Nineteen sixty-one dawned with the inauguration of the youngest President in the history of the United States. John F.Kennedy, JFK, said, "a torch has been passed to a new generation of Americans" and promised action. Within weeks, the White House had become a second Camelot and the charismatic Kennedy and his family seemed to represent a new beginning for America. Sadly, on 25 March 1961, an Atlas ICBM, carrying the first, passengerless, Mercury capsule into orbit, blew up at 35,000 feet. Eighteen days later, the Soviet's Yuri Gagarin was in orbit. It seemed to be the same old story.

Another Soviet first: Yuri Gagarin. (Soviet source)

America's answer: Alan Shepard is fished out of the Atlantic after becoming the first American in space.

In the eyes of the world and of most Americans, America had been beaten again. Fifteen days after this, came the sensational

bungle at the Bay of Pigs and Kennedy's Presidency was beginning to look a disaster. Ten days later, a brighter note was sounded when a Redstone roared into life and Alan Shepard became the first American in space, during a brief 15 minute space hop.

Extremely mindful of the Soviet threat in space and of its effect on the morale of most Americans and of America's image around the world, Kennedy, who was not particularly interested in space, asked his vice president, Lyndon Baines Johnson, what America had to do to beat the Soviets. Johnson, a keen space supporter, told him that a manned moon landing was the only way. As far as Kennedy was concerned, it could have been the launching of anything, anywhere, just as long as it fulfilled the requirement.

So, twenty days after Shepard gave America 15 minutes of manned spaceflight experience Kennedy launched America to the Moon during an historic speech to the US Congress. "I believe that this nation should commit itself, before this decade is out, to the goal of landing a man on the Moon and returning him safely to Earth ..." So it was that the future direction of the American space programme for the next decades, not just decade, was decided on 25 May 1961. The cause of the solid rocket booster malfunction that destroyed the Space Shuttle *Challenger*, killing seven astronauts and absolutely devastating the American space programme and the commercialisation of space, seventeen years after Kennedy's goal had been reached, can be directly attributed to his decision to go to the Moon. So, too, in the same way was his decision to go to the Moon forced on him by the series of events that had followed the launch of the first ICBM in August 1957. Once the pledge had been made, and on the assumption that the Soviets were heading for the Moon too, for America there was no turning back. It was the Moon or Bust. What to do after the Moon was reached was not considered seriously and while getting there Apollo was to overshadow many more directly beneficial applications of space to such a degree that on reaching its goal, most of America asked whether space was worth it. Budgets were cut – leading to a dangerously, cheap Shuttle system – and things were never quite the same again. Perhaps they never will be – unless; unless that unique combination of factors that lead to the Space Race, occur again.

Vice President Lyndon Baines Johnson (pictured in 1968, with Alan Shepard, next to the spacesuit) pioneered Project Apollo and 'sold' it to President Kennedy.

The Media Dream

IF YOU TOLD ANYONE ON 25 MAY 1961 that President Kennedy was an adulterer and that America's first space hero, Alan Shepard had peed in his pants aboard Mercury capsule *Freedom 7* before lift off, you would have been clamped in irons. Certainly, if the press knew, they weren't telling anyone. Kennedy and Shepard were idolized. It was press coverage – that is, the lack of warts-and-all coverage – that played a major role in America's quest for the Moon. The nine years between Kennedy's pledge and the meeting of it were dominated by the cult personalities of the astronauts generated by *Life* magazine and followed by other media, by their brave, pioneering missions and by Moon Race. It was naturally assumed that the Soviet Union was aiming to get there as quickly as possible, too. The Race to the Moon was largely created by our lack of knowledge about the reality of Soviet capabilites and by Western press coverage. And it was this that stirred American efforts and got them to the Moon on July 20 1969.

At the outset, perhaps the most remarkable fact about Kennedy's pledge to get to the Moon in nine years, was that Shepard had gained for America just 15 minutes 28 seconds flight experience, of which only about five minutes was in space. In reality he had merely been up and down on a suborbital flight. No Mercury capsule had even made it into orbit, let alone manned. America had a job on its hands.

While the pioneering Mercury flights continued, America decided how it was to get to the Moon. Straight there and back in one craft? That would require a booster of extraordinary power. Build a two-part spaceship, one part of which lands on the Moon and then rejoins the mother ship in moon orbit? That may not have required such a large booster but it would have meant the mastery of rendezvous and docking in orbit. What could be achieved within the budget and could it do it by 1969? The answer was that to meet the deadline, America would build a two-part Apollo system that was required to perform lunar orbit rendezvous. To practise the skills and rendezvous and docking and to test spacesuits and man's ability to fly in space for periods as long or longer than it would take to fly to the Moon, would be the job of project Gemini. Apollo tests in Earth orbit and deep space would follow, before the actual flight. These would be preceded by unmanned tests and by test flights of the boosters designed to hurl them aloft.

The next nine years in space were to be the most glorious in man's technical history, and perhaps the most exciting period for the public to follow in the press, during a decade that became known, partly through the excitement of spaceflight, as the Swinging Sixties.

How to go to the Moon? This way was too expensive.

Gus Grissom watched *Wyatt Earp* on TV and went to bed. The next day, 21 July 1961 he would become the third man in space. His flight in Mercury capsule *Libertybell 7* was another modest affair, reaching a suborbital height of 118 miles before splashing down in the Atlantic. Minutes later, *Libertybell* was hundreds of fathoms under water and Grissom was ignominiously being hauled, dripping seawater, aboard a helicopter. In truth, he had nearly drowned. No one ever got to the bottom of why the hatch blew off prematurely. "Down from Space and Up from the Ocean" read the headline[1].

On 7 August, America had 30 minutes flight experience, the Soviet Union over 25 hours' worth. Cosmonaut Gherman Titov had just landed after making 17 orbits aboard Vostok 2. Media coverage highlighted the enormous lead, citing the fact that the Vostoks were heavier: "Soviet Space Advance at Each Firing"[2]. Titov had actually been spacesick; the length of flight was to ensure that he landed in a prime recovery zone, which would be overflown only every 17 orbits – or a day – and the capsule was large because the Soviets had not mastered the sophistication of miniaturization. On 13 September, America at last got a Mercury capsule into orbit. But there was only a robot on board. Then, on 29 November another Mercury made two orbits. This time, a chimpanzee, Enos, was on board. That chimps were test piloting Mercurys before the astronauts was a source of great mirth to the real test pilots at places such as Edwards Air Force Base but not to the public and the press. The original intention had been to fly all seven Mercury astronauts on a suborbital flight before sending one into orbit. Then it was decided to limit the suborbital flights to just three. So, before Enos made his flight, astronaut John Glenn had been slated to fly the third mission.

After Enos landed, Glenn was named as the first American to orbit the Earth. His suborbital lob was scrapped and he was next in the queue for the prized mission. Instead of becoming the most anonymous astronaut in history, he become one of the few whose name people remember today. This was the first of many examples of how luck and fate placed astronauts in positions of destiny, including one Neil Armstrong. Glenn's much delayed epic mission, starting with a firey blast-off atop an unreliable and tetchy Atlas intercontinental ballistic missile, on 20 February 1962, was the media's dream, as was the astronaut's personality, reflected in the headlines "Glorious Glenn"[3] and "A Real Fireball"[4]. Although, he had made only three orbits in nearly five hours, it was as if America had beaten the Russians. And Glenn had become the 'first American in space' – Shepard and Grissom had been forgotton.

On 16 March the Titan 2 ICBM, designed to launch the manned Gemini capsules, was successfully tested and on the same day Mercury astronaut Deke Slayton was grounded with a heart flutter. He was to have flown the next orbital mission; indeed, had the Glenn suborbital lob been retained, he would have become the first American in

America's "first" space hero, John Glenn (left), talks to President Kennedy.

orbit. His place was taken by Scott Carpenter, whose *Aurora 7* flight of three orbits on 24 May, was a bit of a shambles. By his own admission, Carpenter was careless. Loss of fuel and late firing of the retro rockets meant a landing overshoot, Carpenter was sighted 56 minutes after splashdown. He lost favour and never flew again. Yet the press coverage of the same event was a little different: "Great Scott"[5], "He's Safe but What a Thriller!"[6]

The most classic case of how media coverage helped to perpetuate the Race to the Moon and placed Russia in the 'lead', was its reaction to the dual flight of Vostok 3 and 4. On 12 August, cosmonaut Andrian Nikolyev was launched. The mission was to be a long one for the first bachelor in space. Then the following day, Vostok 4 was launched. As Nikolyev passed overhead, Pavel Popovich took off. His orbit was different, yet for a brief period the two spacecraft passed within a few miles of each other. Neither craft conducted any manoeuvres. This was not a rendezvous in space. Tell that to the press:

"Spacemen Meet 100 miles up"[7]; "Moscow May Launch Three More This Week"[8]; "I reckon the moon odds 10 to 1 against the Americans now"[9]; "Rocket to the Moon by 1965"[10]. What could Wally Schirra do to follow that? Schirra's modest six orbits in Mercury *Sigma 7* on 3 October was hardly eyecatching, except to Wally himself. "He's down – see page 3"[11].

Wally Schirra's anticlimactic Atlas.

The next astronaut to go was Gordon Cooper who was launched in *Faith 7* on 15 May 1963. "Flash Gordon"[12] or "Gay Gordon"[13], as he was described by the press, made 22 orbits and proved the value of having a man aboard, when he manually controlled retrofire and re-entry. With Slayton grounded and one more flight in the offing, a 48 hour job, Alan Shepard – keen to regain recognition – went into training. His flight was cancelled.

Another media event occurred on 16

June. Cosmonaut Valeri Bykovsky, launched two days before, was 'joined' in space by the first woman in space, Valentina Tereshkova. Vostok 5 and 6 had merely whisked past each other once. But it was another mammoth Soviet triumph, according to western observers. "Space Girl Valya Chases Her Date"[14]. Amateur parachutist Tereshkova had been plucked from the factory floor by the need to perform another space spectacular in Premier Nikita Khruschev's bid to win the Cold War and put the wind up the Americans.

Shortly after this, British scientist Sir Bernard Lovell, who just a year before was proclaiming that a Russian would be on the Moon in 1965, revealed that the Soviets were not interested in getting a man on the Moon but more in establishing a space station in Earth orbit. Tell that to the Khruschev. Suggestions such as these fuelled the first criticism of project Apollo in America. The first budget cuts were made but Kennedy was not around to fight them. The irony of his passing is that his assassination and the manned landing on the Moon were two world events in connection with which most people recall what they were doing at the time.

The testing of the first Saturn rocket designed to carry Apollos into earth orbit for test flights occurred on 30 January 1964. On 13 April, the first Gemini was launched on an unmanned orbital mission, and the first two-man crew to fly the spacecraft was named. They were Gus Grissom and crew-cut, 33-year-old Lt Cdr John Young, as quiet an astronaut as could be imagined. The first 'boilerplate' Apollo was launched into Earth orbit on 29 May. On 29 July, James McDivitt, chosen as commander of the second manned Gemini flight, said that walking in space would be a highlight of the mission the following year. The following day, at last, after six mishaps, *Ranger 7* took the first close-ups of the Moon surface. "Man CAN land on the Moon"[15]. On 18 September, an unmanned Apollo command service module was placed into orbit by a Saturn 1.

The Soviets had not launched a cosmonaut for nearly 18 months. Things were hotting up in the USA. Had the Russians been left behind? October 12: "Russian Three Man Spaceship in Orbit"[16]; "Commander, Engineer and Doctor have Lunch Together"[17]; "Two Year Lead by Russians in Space Race"[18]. Was Voskhod 1 really a three-man spaceship? No, it was actually a one man Vostok with three men crammed inside, without spacesuits and with no means of emergency ejection, making it the most perilous manned mission in history. If the Americans are going to launch two men, we'll launch three, said Khrushchev. Ironically, he was removed from office while congratulating the three 'pioneering' cosmonauts.

Five days before America was due to launch Gemini 3 on the first manned mission in the programme on 25 March 1965, the Russian launched Voskhod 2. This time there were only two men on board. The reason become abundantly clear shortly after when the heavily spacesuited Alexei Leonov crawled through an airlock, which had replaced the third seat, and walked in space. "Russian Makes History With Walk in Space"[19]; "Now it is Chase to the Moon"; "Barnum and Bailey never thought of this. Colonel Alexei Leonov's space thriller is the greatest circus act in the history of the Universe"[20]. How do the taciturn twins, Gus Grissom and John Young follow that? "Space drive by Two Quiet Americans"[21]. Gemini 3 made only three orbits but Grissom, using the first space computer, changed the orbit in the first step towards rendezvous and docking.

On 3 June, Gemini 4 took off and the upstaged Edward White made the first American walk in space, using a hand-held manoeuvring unit. "Ed takes a space walk"[22]; "Jay walk in space"[23]. Photographs of his exploit are some of the best ever taken and are a credit to commander James McDivitt. By 7 June, America with 62 orbits, was at last on par with the Russians, according to the press and observers. This was a turning point. America never looked back and chalked up an extraordinary string of space successes, launching eight more Geminis by November 1966 by which time the last Russians in space were still Voshkod 2's Leonov and Pavel Belyayev.

First, Gemini 5, with Gordon Cooper and Pete Conrad aboard, made the first on-time manned lift off in the US space programme on 21 August. The spacecraft was designed to rendezvous with a small target craft ejected from the nose of Gemini, which also carried hydrogen-oxygen fuel cells to generate electrical power for the first time. These malfunctioned, prompting the headline "Crisis in Space"[24]. With power low, the astronauts were left to drift through space, not doing much more than looking out of the window. That would have been fine for a day but they were going to be up for eight. Conrad described it as a very boring flight. But by staying up for eight days – the duration of a Moon flight – it was a case of "Now for the Moon Shot!"[25]. A month later, at the International Astronautical Federation congress in Athens, Wernher von Braun, now a leading force behind Apollo, was confidently saying that Americans would fly to the Moon in 1968. They wouldn't land, he said but just fly around. At least it would be sure to beat Russia.

Testing the crucial ability to rendezvous and dock in space was to be the job of Gemini 6, to be launched on 25 October. Dependable Wally Schirra and balding Tom Stafford were to perform the extraordinarily complicated pirouettes in space that would enable them to rendezvous with an Agena target rocket. The rocket, however, blew up and Gemini 6 was stranded. "Space Shot Hitch Drama"[26]. Gemini 7 was planned as a purely long duration mission, during which two astronauts would live for 14 days inside a cabin no larger than the front seat of a Volkswagen. On 29 October, President Johnson announced that Gemini 6 and 7 would perform the first space rendezvous.

As the Russians were trying unsuccessfully for the fourth time to soft-land an unmanned Luna spacecraft on the Moon, the great adventure began. Gemini 7 was launched first, on 4 December. Gemini 6 was to follow on 12 December. All went well, the Titan's engines emitting the now-familiar high pitched whine and igniting. The spacecraft clock started. Lift-off. The engines were shutdown. Gemini 6 had not lifted off. An all-rookie crew might have been misled by the operating clock into thinking that they had lifted off and, as a result, could have yanked the ejection levers. But the veteren Schirra earned a new nickname, 'Mr Cool', by staying put and saving the mission. The headline "16 minute rocket peril"[27] summed up the anxious moments as the astronauts waited to get out and technicians feared an explosion, because the hypergolic propellants of the Titan would ignite spontaneously on contact. This was to be one of Schirra's two major contributions to getting Apollo 11 to the Moon on time. Things went swimmingly on 15 December. The greatest hurdle on the road to the Moon thus far, had been achieved.

"Space Four Waltz Cheek to Cheek"[28]; "Howdy Spacemates"[29]. "Just Six Feet Apart

Ed White on EVA from Gemini 4.

The greatest milestone: Gemini 6 and 7 nose to nose.

in Space"[30]; – in fact, they came to within one foot of each other. By 18 December, everyone was home and the health of 'Itchy and Crummy' Frank Borman and James Lovell, as they described themselves, was remarkably good after two weeks in a can.

Although no cosmonauts were launched in 1966, the Soviets did manage to hard-land the tiny Luna 9 on the surface of the Moon and return TV pictures. These gave the Americans confidence; astronauts weren't about to sink into thick moondust. At the time, few in the West, it seemed, realized that Luna was dropped like a stone from a crashing rocket stage and bounced along the surface until it came to a standstill. Nor did they realize that it couldn't hold a dog, let alone a cosmonaut. So much so, the Sir Bernard Lovell was admiring the skill of the Russian's soft landing and claiming that a Russian could be on the moon by 1970. The enthusiastic and entertaining British TV astronomer-cum-'space expert', Patrick Moore, was similarly fooled later and cut off an argument with this then 18-year-old stripling, at a meeting of an astronomical society, by mocking my source of 'sour grapes American information' about Luna 9: actually an *Aviation Week* report based on Russian information! On 28 February, the two men intended to crew Gemini 9 were killed when their jet hit the building in which Gemini 9 was housed, bounced off and exploded in a car park. Their place was taken by the back-up crew which had been flying behind them but which landed safely. Although Gemini 6 had performed the rendezvous, it was Gemini 8's task to make a docking with an Agena target rocket on 16 March. The command pilot, Neil Armstrong, made a typically neat, no-nonsense job of this, even docking out of radio contact with ground control. Later, he and his co-pilot were lucky to escape with their lives when a thruster on Gemini 8 short-circuited and fired continuously, sending the Gemini and Agena into a violent 70 rpm spin. Untrained men would probably have blacked out. Control was saved and an emergency splashdown followed. "Splashdown SOS after link up fails"[31].

Before Gemini 8, the first Saturn 1B launch of a fully recoverable Apollo command module had taken place and, after Gemini 8, NASA named the three man crew to fly the first mission. It was to be Gus Grissom, Edward White and new boy Roger Chaffee.

In bouyant mood, NASA set to work on Gemini 9, which would repeat Gemini 8's task. An exploding Agena target rocket and a launch cancellation at T+2 minutes meant that Tom Stafford had been to the pad four times for just one lift off. But on 3 June – the day the US Surveyor 1 made a true rocket-aided soft landing on the Moon – Stafford got off the ground and into trouble. The alternative target craft couldn't allow dock-

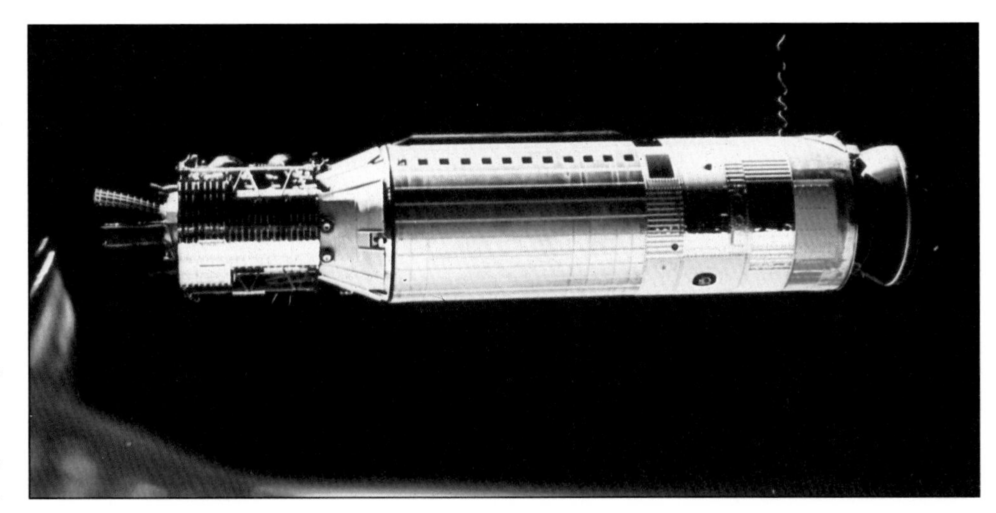

Neil Armstrong's first target, Agena 8, before the great spin in space.

Spaceport USA. Gemini 9 takes off with a Saturn 5 in the background.

ing because its payload shrouds had not parted. "Space Date with an Angry Alligator"[32]. Then Eugene Cernan's spacewalk was called off when his spacesuit's environmental control system could not cope with the sweat and heat he was generating as he worked outside. "Walking in Space Fog"[33]. Cernan had planned to fly a tethered manoeuvring unit. But at last, something went perfectly with a "Bullseye"[34] splashdown.

By this time, the first Saturn 5 booster, all 363 feet of it, including a mock-up Apollo on the top, had been rolled out of the biggest building in the world and to the launch pad. Things were going well. There were even discussions within NASA of flying Apollo 1 on a rendezvous mission with Gemini 12 and sending Apollo 2 around the Moon in 1967. "The success of Gemini and Surveyor and the slowdown of the Russian effort, is encouraging a more audacious mood"[35].

"The world is round!" exclaims Gemini 11's Pete Conrad.

Geminis 10 to 12 were to be the icing on the cake. First, the taciturn John Young and equally quiet Mike Collins were launched on 10 July, to dock with Agena 10, relight its engine and soar to a record height of 475 miles, rendezvous with Agena 8 and retrieve samples from its side. It seemed a breeze, despite a halted spacewalk prompting the headline "The High Life for Spacemen"[36]. Meanwhile, the unmanned Lunar Orbiter was taking some close-ups of the Moon in preparations for choosing the moonlanding site and, for the very first time, photographing the Earth from the Moon. On 12 September Gemini 11 took off. Ninety minutes later it had docked with Agena 11, simulating the ascent of the Apollo lunar module and rendezvous and docking with the command module: "Moon trip trial"[37]. The Agena 11 engine lit up, taking Pete Conrad and Richard Gordon to a record earth orbit apogee of 850 miles. But Gordon's spacewalk hit the perennial problem: "Blinded Spaceman Calls Off His Walk"[38]. Using a tether attached to Agena 11, Gemini also created some artificial gravity before the astronauts folded their arms for an automatic landing. "Look No Hands as Space Crew Splashdown"[39]. Even rumours that Russia was about to launch a rocket larger than Saturn 5, in a last bid to beat the Americans to the Moon, did not dampen the spirits as Gemini 12 took off on November 11. Even Aldrin's spacewalk was a great success: "Buzz Makes Two Hour Walk Around the World"[40]. Then it was, "Final Splashdown for Gemini – now US looks to Apollo and the Moon"[41].

Things were building up to a crescendo in December 1966 when Lunar Orbiter 2 returned an incredible picture of the crater Copernicus from just 28 miles up. "The Picture of the Century" said NASA. Apollo 1 was set for blast-off on 21 February 1967. The mission was for a 14-day shakedown in Earth orbit, launched by the smaller Saturn 1B, but beneath the high public confidence there was concern about the Apollo capsule. Apollo 1 was behaving so badly during ground tests, that after one, a disgruntled Grissom had put a lemon on its top. Then came the shock. Late news: "Astronaut feared dead in fire"[42]; 2am news: "Three spacemen die in blast"[43]; Last edition: Death Trap Riddle of the Moon Men"[44]; and "The last shout: Fire in the Spacecraft!"[45]. Shocking overconfidence had bred bad quality-control and design. Flammable materials in a pure oxygen atmosphere, a short circuit, a spark, a fire, lots of choking smoke, an inferno and a final surge of flame as the spacecraft burst. Grissom, White and Chaffee choked to death, but it has never been revealed just how badly their bodies had been burned.

Nineteen sixty-seven was the year that America came to a stop. However, but for the imperfections of Apollo that were exposed so cruelly by the fire and the modifications and improvements made to Apollo afterwards, it is doubtful that Apollo 11 would have reached the Moon when it did.

Russia re-entered space in 1967. Soyuz 1 was launched on 23 April with Vladimir Komarov on board. Soyuz was a spacecraft designed to link up to form a mini space station, as a ferry to fly to a larger space station, and as a manned lunar looper. Russia still had a chance to beat the Americans to the Moon, albeit not with a landing. Soyuz 1 was to have linked up with a Soyuz 2 but never did. "Death Dive From Space – Capsule Parachute Ropes Tangle"[46]; "Russian Dies in first Space Crash"[47]. Russia had been grounded, too.

People were being reminded of the perils involved in exploring space. Calls were made for a joint US-Soviet flight to the Moon. "Now, maybe they'll stop this senseless Moon RACE", wrote the *Daily Mail's* Angus McPherson[48]; "Let's link up for Moon Bid"[49].

The year of 1967 was a tragic one, indeed. With little publicity NASA astronaut Ed Givens was killed in a car crash, colleague C.C. Williams – who would have become the fourth man on the Moon – in an air crash, USAF Manned Orbital Laboratory astronaut Robert Lawrence, the first negro to be chosen was killed in an air crash and X-15 'astronaut' Mike Adams was killed on a mission.

By 9 November it was all systems go again, as the first Saturn 5 booster was launched. The noise and vibration of the blast-off shocked onlookers. Walker Cronkite, the CBS anchorman describing events at the Kennedy Space Centre, thought that his studio roof was going to cave in. The headlines read, "Saturn Spectacular"[50]; "Away She Goes"[51]; "On Our Way to the Moon"[52]; "US Races to the Moon"[53]. The following January, 1968, the Apollo Lunar Module was launched by a uniquely-configured Saturn 1B. Another success: "The Bug Beats Last Barrier"[54].

Remarkably, the first manned Apollo flight hadn't yet taken place. Before it did, the Russians launched Zond 5 on a lunar looping flight and recovered it safely: "Moon Lassoed By Russian Space Ship"[55]; "Moon Success gives Russia Space Lead"[56]. A Zond flight with one cosmonaut was expected by December. As Apollo 7 prepared for its maiden voyage, there was again speculation about Apollo 8 going to the Moon – and into lunar orbit. Apollo 7 was launched on 11 October: "Mr Cool takes his rookies into Space"[57]; "We're Having a Ball!"[58]. Schirra, Donn Eisele and Walter Cunningham performed well, amid bad moods and high jinx. Schirra, who had a cold, behaved like a prima donna but was pretty cheerful during in-flight TV shows: "Spaceman Blues. He's One Degree Under in Space"[59]; "Watch this Space For Laughs!"[60]. Apollo 7 was a highly successful shakedown flight and, as expected, Apollo 8 was aimed towards the Moon. Before it, however, Russia launched a red herring mission, called Soyuz 3, which did not go to the Moon nor dock with the unmanned Soyuz 2. "Russian in Orbit – a Moonshot?"[61]. The Soviets ran out of time and on December 21 the epic mission of Apollo 8 began with men riding only the third Saturn 5. "Out of this World"[62]; "A-OK For Rings Around the Moon"[63]; "Lunarnauts – and Spot On!"[64]; "Final Peril"[65]; "Masters of the Moon"[66]; "And now for Apollo 9, 10 and TOUCHDOWN!"[67].

'In the beginning. . . .' from Apollo 8.

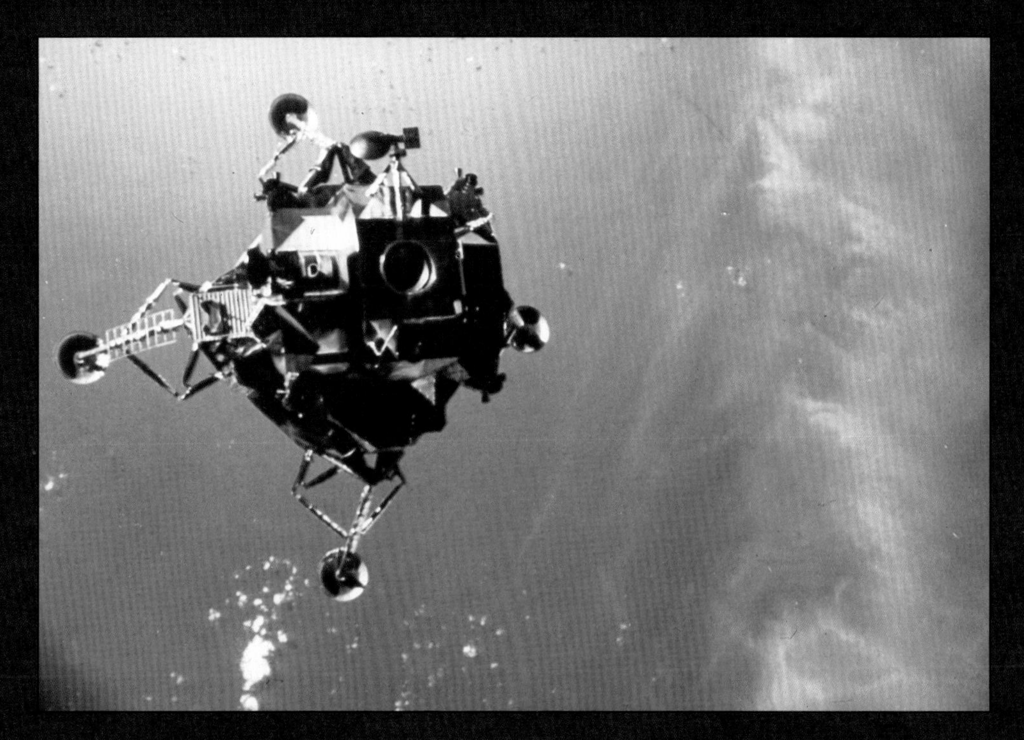

Apollo 9's manned Spider *lunar module in Earth orbit.*

Snoopy comes home to Apollo 10's Charlie Brown *and all is set for Apollo 11.*

What had been described as 'Moon Flight Madness' by Sir Bernard Lovell, captured the imagination of the world, particularly as it happened over Christmas. Few can forget eating their Christmas dinner, listening to the voices of astronauts Frank Borman, James Lovell and Bill Anders reading passages from Genesis and ending with good wishes for, "all of you, all of you on the Good Earth". But Apollo 8 was, ironically, taking place amid budget cuts which were cancelling later landing missions.

On 9 January 1969 the crew of Apollo 11 was named: Neil Armstrong, Mike Collins, Buzz Aldrin. It may have been the year of the Moon but Russian ventures still diverted attention. Soyuz 4 and 5 docked in Earth orbit and there was an external crew exchange: "Knock, Knock, Who's There? .... Space Walkers!"[68]. Apollo 9's Earth orbital checkout of the Lunar Module on a 'lunar landing mission' began on March 3. "Moonbug is all set for its Solo in Space"[69]. At the same time, President Nixon was authorising more budgets cuts which were threatening Apollo just as its climax approached. "The fuse that burns slowly away at Cape Kennedy"[70], intimated that time was running out. The lunar module Spider undocked and flew its 'landing and take off' in Earth orbit. It was the first manned flight of a craft which could not survive re-entry. The astronauts, Jim McDivitt and Rusty Schweickart, had to dock with Apollo command module Gumdrop. "On Their own The Spidermen Spin in Space"[71]; "Do or Die Space Dive"[72]; "Whew! Apollo Men Make It"[73]; "Goody, Goody Gumdrop"[74]. James McDivitt, David Scott and Rusty Schweickart flew a perfect mission. "US Set for Summer Moon Trip"[75].

As preparations progressed for the Apollo 10 mission to do a repeat of Apollo 9, but in Moon orbit, there was speculation about just who would be the first man to set foot on the Moon. Would it be Apollo 11 commander Armstrong or lunar module pilot Aldrin? "Astronaut Edwin Buzz Aldrin, world champion spacewalker, is expected to be the first man on the Moon. That's the plan said John Small of NASA's space centre". (12 March[76]); "Neil Armstrong, the civilian commander of Apollo 11, is likely to be the first man on the Moon, says George Low of NASA". (14 April[77]).

As 18 May approached the build-up was getting intense but still to be endured were the dramas of Apollo 10 or the boredom depending on whether you were hooked on the Moon. "Riskiest Space Flight Yet"[78]; "Last Rung on the Ladder to the Moon"[79]; "What a Ride. It's Fantastic, Babe"[80]. Tom Stafford and Eugene Cernan got to within nine miles of the surface – and into trouble: "Moon Buzzed, then Trouble"[81]; "That was wild baby!"[82]; "Perfect Splashdown Crowns Success of Apollo 10 flight"[83].

After weeks of endless press coverage about the Apollo 11 mission and its crew, 16 July approached. The *Daily Mail's* ace space correspondent, Angus McPherson, summed-up eloquently the feelings of the world: "There is no more time. The green, twitching figures of the countdown clock run out like grains of sand. In the sky the crescent of the new moon grows like the first chink of a slowly opening door. Tomorrow, three men will set off towards that door and all the unknowns that lie behind it. In the 470-man control room here the last call before the gleaming white skyscraper, the Saturn 5 rocket, heaves itself up off Pad 39A will be 'commit'. Neil Armstrong, Mike Collins and Edwin Aldrin will be committed indeed. And so will we."[84]

Apollo 11

33rd manned spaceflight
16 July 1969
8 day 3 hr 18 min 35 sec
21 hr 36 min on Moon
48.5 lb of Lunar samples

The First Steps at Tranquillity

SUCH WAS THE EXTRAORDINARY BUILD-up to Apollo 11, that pictures and stories about the crew were appearing on the front pages of newspapers and magazines around the world weeks before lift-off. The air of confidence that Apollo 11 was 'The Beginning' was illustrated in predictions that, following ten Apollo landings on the Moon by 1972, there would be manned flights to Mars in 1985. People were being told that there had been nothing like it before – and perhaps there will never be again. This was more than just the highlight of space exploration, we were told. But there was still, maybe, a feeling of slight anticlimax. After all, hadn't there already been two highly successful manned Moonflights?

Part of the media build-up was the crew's pre-launch press conference, which was shown live on TV on a Saturday evening in Britain, without viewer complaint. During it, Armstrong came over in a strange, halting manner, drawling the names of craters he would fly over towards the Moon's surface. Aldrin was the epitome of clipped, military precision and Collins, naturally, raised the laughs. The crew attended the press conference behind a glass screen, after entering the room wearing masks – testimony to the exhaustive efforts being taken to prevent their catching Earth bugs, rather than Moon bugs. Asked what precautions he had taken at home to prevent catching anything untoward, Collins told the press, "My wife and children have signed statements that they have no germs". President Nixon decided he would defy the bugs and attend the pre-launch dinner with the astronauts. Astronauts' doctor Chuck Berry refused to allow it and Nixon sent his regrets. This rebuff probably made Nixon even more determined to share the limelight – on the Moon.

Nixon had already ensured that his name would be preserved for posterity, by including his signature on the plaque that would be delivered to the surface of the Moon on one of the landing legs of the Apollo 11 lunar module *Eagle*. An eagle was to be the subject of the mission badge, which did not include the crews' names. The claws of the eagle seemed a bit threatening, so an olive branch was added. With Jules Verne in mind, the command module was named *Columbia* by the crew.

The crew were certainly laying their lives on the line and were more than aware that things could go wrong. This, however, was not alluded too very much by the media, except perhaps of the prospect of Armstrong and Aldrin being stranded on the lunar surface with about 42 hours worth of of air, if Eagle's engine failed to lift them off it. Armstrong commented before the flight that, yes, of course he was aware but it was "something we prefer not to think about". According to Christopher 'Columbus' Kraft, a

mission control chief, 'it doesn't happen many times in a man's life that he conciously and deliberately faces death and these men are putting their lives on the line. Everybody is aware of it. The crew is trying to do something no man has done before.' If truth be known, Armstrong had already been laying his life on the line every time he boarded the grasshopper-looking lunar landing trainer, which had a chequered history. During one malfunction, Armstrong ejected, only to find himself falling into the burning debris of the vehicle. The prevailing wind saved him that time.

Launch of Apollo 11 as seen from the pad tower.

Right: The Sea of Tranquility. Apollo 11's target area.

Far Right: Armstrong, already on the surface, takes picture of Aldrin, emerging from the lunar module.

Right: One of Armstrong's fine photos, showing Aldrin with the flag.

To add spice to the pre-launch nerves and tension, the Soviet Union launched Luna 15, an unmanned spaceraft. This was scheduled to soft-land on the Moon, take a very small sample of the lunar dust and return it to Earth before Apollo got home.

15 July: "Message to the waiting world: We have No Fear"[1], said the astronauts. Armstrong reckoned that the flight had an 80 per cent chance of success, but that if it did fail, then he'd be aboard Apollo 12, promised NASA administrator Thomas Paine. What if Apollo 12 failed too? 16 July, launch day: "It is at once so inconceivable and so familiar – a journey to the Moon – that the event becomes blurred in unreality"[2].

Over one million people flocked to vantage points near the Kennedy Space Centre and over two thousand pressmen from all over the world milled around the grass area in front a press grandstand, which was housing the 'working' media, reporting live to radio and TV stations. One of them was the British veteren Reg Turnill of the BBC, who had watched every manned spaceshot since Alan Shepard. Another was the more widely known Walter Cronkite of CBS, which had its own studio at the space centre. Cronkite had even held a 'Meet Walter Cronkite' cocktail party at one of the many Cocoa Beach hotels. VIP guests watching

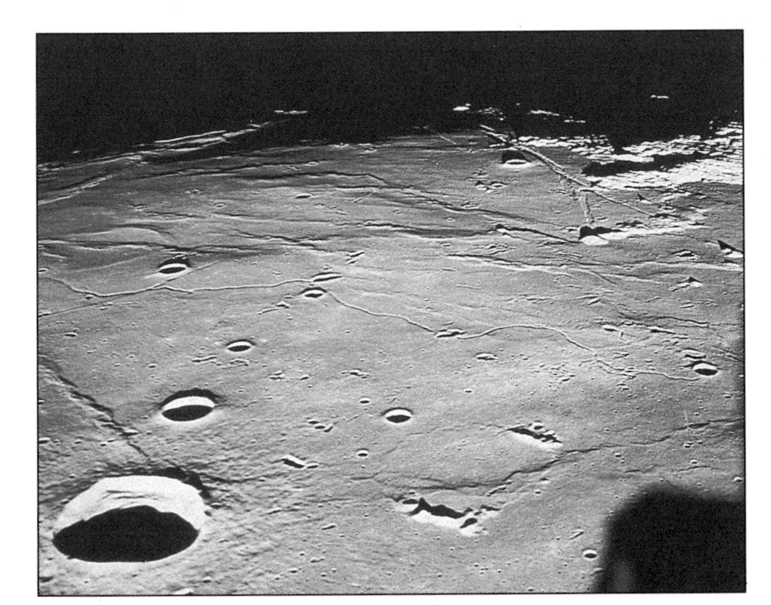

nearby included all the US congressmen and senators, who had supported or not supported the space programme, ex-President Johnson and his wife Lady Bird, who had, and vice president Spiro Agnew who was to be pictured watching the blast-off, looking skyward with a glass of bourbon in his hand. More than 1,000 million people were watching the event live on TV.

From the press site, the gleaming white, 363 feet monster Saturn stood in bright summer sunshine about three miles away. A louspeaker blared the countdown reports. As 2.32 pm Kennedy Space Centre Time approached, the tension and excitement rose to fever pitch. This was it. The serene looking Saturn, gently puffing clouds of vapourising liquid oxygen propellant, would soon be transformed into a second Sun. "Twelve, eleven, ten, nine, ignition sequence start. . . ." A huge ball of reddish-orange flame appeared at the base of the Saturn. "Six, five, four. . . ." Huge clouds of steam gushed out from the left and right of the pad as the ball of flame became incandescent. "All engines running, three, two, one, zero, lift-off. We have lift-off, thirty two minutes passed the hour. Lift off of Apollo 11. . . . tower cleared." To the cheers and cries of observers, the Saturn rose majestically towards the blue sky – unbelievably in total silence. Then it came. The noise. The almost indescribable ear-shattering, ground-shaking roar and whiplash crackles of shock waves. In a famous *Life* magazine photo of the scene at the press site, one of the observers is seen walking away, with his back to the rising Saturn, the only person in the area, perhaps, who wasn't actually watching the event, except the crew inside the command module.

Armstrong, seated left, Aldrin, in the centre hatch and Collins on the right, did not see a thing out of their windows for the command module was covered with a blast shield from the launch escape tower, which would be jettisoned, along with the shield, later in the ascent. Minutes later, after brief comments from the quiet crew, Armstrong said, "Shutdown" as the third stage stopped. Apollo 11 was in Earth orbit. "The overture to a new era of civilisation", said Spiro Agnew supping his bourbon. "2.32 on a summer day . . . and for good or evil, a new beginning"[3]. Remembering Luna 15, one newspaper said, "Race for the Moon"[4]. Another reminded us of the cost: "It has cost America something like $10,000,000,000 to reach this point"[5].

The S4B ignited again two orbits later, sending Apollo 11 on its translunar coast. Collins, having moved into the centre seat, gingerley jockeyed the *Columbia* command module into the open shrouds of the S4B and extracted *Eagle*, the lunar module. Soon, the TV camera came out and the quiet but friendly crew showed viewers what the Earth looked like out of the window. The equally friendly astronaut Charlie Duke, the capsule communicator specifically requested by Armstrong to be on duty during the moon landing phase, asked Armstrong if he could "turn the Earth a little bit so we can see a little less water?" The lighthearted banter and leg pulling of Charlie Duke was a feature of the pre-Moon shows, which were included in the many TV 'Moon Specials', In Britain these were introduced by the ebullient chatterbox James Burke of the BBC, who had the knack of talking as soon as the astronauts said something, and the somewhat sheepish and softly spoken Peter Fairley of ITN, whose Apollo productions were masterminded by Frank Miles, the company's head of science. Another feature of the translunar coast (TLC) was how the astronauts were being kept in touch with how the press was covering the flight. Collins was told that Armstrong was described by the Soviet Union's *Pravda* as the 'Czar of the Ship'. Later Collins commented dryly: "the Czar is brushing his teeth". Meanwhile, the Russians were also assuring astronaut Frank Borman, who had called Moscow, that Luna 15, now in moon orbit, would not interfere with Apollo 11.

While the crew busied themselves in not-so solitary space, their wives had to undergo the statutory media ordeal, which included standing outside one of their houses with the Stars and Stripes as a backdrop, being asked stupid questions and answering diplomatically; at one point, even, saying the statutory words in unision, "We're proud, thrilled and happy". The wives were the Apollo 11 ground crew; in support were other astronauts, who would translate the technological gobbledegook coming from Apollo into squawk boxes installed in each home, to reassure them. Most of the crew conversations, in fact, were impossible to undertsand to those outside the space programme. "TEI, 11, SPS, G&N, 37,200 minus 060 plus 047 plus 098," says a voice from Apollo. "Roger 11," says the familiar Duke.

"It was like perfect," said Armstrong as Apollo 11 appeared from the far side of the Moon, confirming that the *Columbia-Eagle* combination was in orbit. Sunday 20 July, and British newspapers heralded the moon-landing:

"Now for a Bit of Rock!"[6]; "Man lands Today. Off to Work – in water-cooled underwear, and it's British by jove"[7]. Luna 15 was still a worry: "Has the Red Robot by a Short Head"[8]; "Today, after a million years on planet Earth, Man lands for the first time on another world"[9].

On Sunday evening the TV specials started the big session – The landing. That the landing came within seconds of an abort and was saved by a computer engineer at the Houston control centre was not apparent at the time. As *Eagle* was on its powered descent initiation (PDI) burn its computers were working overtime and protesting. Alarms kept sounding in *Eagle's* cockpit. In the Houston control centre, Steven Bales' job was to monitor the computers and he decided that they could be overidden. He told the flight controller, who in turn told the desperate-sounding Duke, who shouted to *Eagle*, "Hang tight, you're go!" With thirty seconds of fuel left, *Eagle* was still airborne, its engine churning up dust as Armstrong, having taken manual control, steered for a smoother area than the rock-strewn crater *Eagle*'s autopilot was taking it into earlier.

The engine stopped. The dust settled. The first words from a man on the Moon were "Engine stop", as Aldrin shutdown various systems. "We copy you on the ground," said Duke. There was a deathly hush. Armstrong, with impeccable, dramatic, theatrical timing, uttered the official first words: "Houston, Tranquillity Base here. The Eagle has landed". Charlie Duke's response also deserves a place in history because it is almost always heard when the landing recording is replayed. He obviously had not expected Armstrong to describe his landing site so formally. The breathless capcom was also extremely relieved, having oozed almost all of his adrenalin over the landing. "Rog . . . Roger Twan . . . Tranquillity. We copy you on the ground. You've got a bunch of guys about to turn blue. We're breathing again. Thanks a lot!". There was pandamonium in mission control, TV studios and living rooms. This was one of the greatest moments in history, heard live as it happened. Armstrong and Aldrin, never the closest of friends, merely shook hands with each other. Before he performed a religious ceremony on board

Eagle in thanksgiving, Aldrin asked the listening world to reflect silently.

The original plan had the two astronauts sleeping before the moonwalk but, not surprisingly, this proved difficult and the excited duo donned their portable life support system backpacks in the tight confines of the lunar module and prepared to exit. Meanwhile, Mike Collins, like John Young before him on Apollo 10, was the most solitary man in the Universe as he orbited the farside of the Moon. "Don't forget the one on the command module," he would say later.

Britain waited through the night to witness the first footsteps on the Moon. Even the TV specials shut down for a while. On Monday morning, 21 July in Britain – 20 July, of course, in some other parts of the

Aldrin takes the camera and, according to the flight plan, photographs his boot and boot print.

According to the flight plan, Aldrin takes a 360 degree panorama photograph. Armstrong happens to be in the picture. No formal still picture of Armstrong on the surface exists because Aldrin didn't take any. (Picture supplied by H. J. P. Arnold of 'Space Frontiers')

world – millions of TV viewers watched an event that they will never forget; as with Kennedy's assassination and personal family events, especially bereavements, this was an event during which people will always remember what they were doing at the time. Britain had a choice, the BBC with Cliff Michelmore, Patrick Moore and Burke, with his pointer at the ready, or ITN with the urbane and avuncular Alastair Burnett and quietly enthusiastic Fairley.

The wait for something to happen became nearly intolerable when, at last, *Eagle*'s porch door opened and Armstrong on his hands and knees crawled out backwards. As he slowly came down the rungs of the ladder, he turned on a TV camera. The black and white picture crackled and an apparition appeared – upside down. The picture righted itself and the irrepressible James Burke could contain himself no longer, pointing at the obvious spacesuited figure and saying emphatically, "There is Armstrong!" What Armstrong did and said after this is well documented. Or is it? History books will say that he said, "That's one small step for a man, one giant leap for mankind." Anybody listening will say he said, "That's one small step for man". Armstrong maintains he said 'a' man. When did he decide what to say? Armstrong says not until after the landing.

A little deadpan humour ensued when Aldrin exited and closed the hatch slightly, commenting as he did that he was just "making sure that I don't lock it on the way out". When he stood on the Moon he said quietly, "magnificent desolation". The US flag was deployed on the Moon, while Aldrin posed for many of Armstrong's superb

pictures. Only once did Aldrin take the camera and take a picture of Armstrong, when he had his back to the camera and was in the shadow of *Eagle*. This took place during a planned 360 degree sweep of panoramic photos by Aldrin, who was never instructed in the flight plan to take pictures of the CDR. Nevertheless, Armstrong took many unscheduled photos of Aldrin.

Armstrong, meanwhile, clearly had received a warning that the brief exploration of the surface of the Moon would be interupted by the gushing, over-the-top Nixon. "Hello, Neil and Buzz. I am talking to you from the Oval Office at the White House and this has got to be the most famous telephone call ever made. . . ." Armstrong's reply was eloquent.

President Nixon gets involved again, with the astronauts a captive audience inside their quarantine container on the USS Hornet.

Scientists were besides themselves. Enough of the ceremonies. Get to work! This had included Armstrong taking a contingency sample and putting it in his spacesuit trouser leg in case they had to take off quickly. Said fellow astronaut Don Lind, "He's going back in with his pants on, so we will have that sample for sure". As Britain went to work on Monday morning, Armstrong and Aldrin returned to the cramped confines of *Eagle*, after the latter had performed some comical looking Kangaroo hops.

Newpapers had predictable headlines that morning. However the headlines were larger than usual.

"Man lands on the Moon with perfect touchdown"[10]. "ITV stays silent and beats chatty BBC"[11]. Playing the Moon race to the end, one newspaper had "Americans First on the Moon"[12]. Luna 15, meanwhile, was to crash in the Mare Crisium. "Dateline. Tranquillity Base – Moonday"[13]. "Today these three men lead mankind into a new world . . . and a new era"[14]. "Moonfall. The most fantastic voyage in human history"[15]. And, more thoughtfully: "Tonight. Moonmen face new peril"[16].

Stranded on the Moon. This was the greatest fear about Apollo 11. Would the small engine of *Eagle*'s ascent stage ignite? If not, the astronauts had precious little time to attempt a fix. "Fear is not an unknown emotion," Armstrong had remarked before the flight. If there were any fears, they were not obvious as *Eagle* prepared to take off. "Understand, we're number one on the runway," said Aldrin, sounding jocular at last.

Following a crackling ignition, *Eagle* rose rapidly into the skies above Tranquillity, which now looked like Brighton beach after a summer afternoon, leaving it in chaos as the exhaust blew down the flag and caused a dust storm among the debris the crew had left behind together with the science instruments. The *Sun* commented that *Eagle* left behind, the most expensive junkyard in the Universe, including a TV camera, Hassleblad cameras, portable life support systems, tongs, scoop, hammer – and the descent stage of the lunar module[17]. The crew had also left behind a microdot message from leaders of the world, a memorial to fallen astronauts and the crew patch for Apollo 1 – Grissom had got to the Moon after all.

Eagle performed a perfect rendezvous and docking with Columbia – until the last moment when there was a scary juddering. "Nice to have some company at last!" said Collins. The *Daily Mirror*'s Brian Hitchen predicted, with errie accuracy in the case of Aldrin, that "The tremendous pressures they face during their lunar journey will seem small compared with the social and financial pressures which will be forced upon them when they back to Earth"[17]. Astronaut Frank Borman was to tell the crew after they returned, that they need never worry financially again. "Gee, thanks Frank," replied Armstrong, possibly surprised that someone like Borman would make such an crass remark.

During the final TV broadcast from Apollo as it was heading home, having discarded the ascent stage of the lunar module in lunar orbit, the three crewmen gave their reflections on the epic mission. Each reminded the world that Apollo 11 was a team effort by thousands of Americans, a fact so easily forgotten. Not easily forgotten was the fact that waiting for their return was President Nixon, on board the recovery ship *USS Hornet*, latest speech in hand. Also on board was the quarantine container in which the crew had to travel during their journey to the lunar receiving laboratory at Houston, in which they would have to stay for three weeks, to ensure that they and their moon-rocks brought no Moon bugs home with them.

The TV programmes were looking a bit

tired by now and coverage of previous Apollo splashdowns and recoveries had been somnolent affairs because the recoveries took so much time. All the viewers wanted to see was the helicopter landing on the carrier deck and the jubilant crew hopping about. But on Apollo 11, they didn't even get this. The crew had to don biological isolation garments as soon as the spacecraft hatch opened – after an upside down splashdown – and were faceless individuals as they were pictured leaving the helicopter to enter the container. A little later the three appeared by a window of the container after opening a twee little curtain. Collins sported a moon-grown moustache. They looked out and there was Mr President. "I want you to know that I am the luckiest man in the world, not only because I am honoured to be the President of the United States but because, particularly, I have the privilege of speaking to you for so many in the world welcoming you back to Earth. This has got the be the greatest week since the Creation because as a result of what has happened this week the world is bigger, infinitely. As a result of what you have done the world has never been closer together and we should thank you for that."

The container was duly delivered to Houston and it wasn't until 10 August, that the three men were allowed to leave and go home to their wives and families. In no time at all, they were in such a constant whirl of banquets, press conferences, dinners, speeches that they soon forgot where they were and what time it was. The three had not expected this, nor had they been trained for it. There then followed a staggering 25-country world tour, between 29 September and 4 November, in President Nixon's jet to make as much political capital out of Apollo 11 as was possible. The irony in this was that not only had Nixon cut the space budget after inheriting Apollo from President Johnson but a Kennedy no less, had just stolen Apollo 11's thunder – by driving off a bridge at a place called Chappequidik.

Apollo 12

37th manned spaceflight
14 November 1969
10 day 4 hr 36 min 25 sec
31 hr 31 min on Moon
74.7 lb of Lunar samples

Thunderstruck

en route to the Ocean of Storms

AS THE APOLLO 11 EUPHORIA WAS apparently wearing off, 35,000 people descended on the Geological Museum in London to see 10 grammes of moondust, enough to fill a couple of desert spoons. Wernher von Braun was suggesting that the President of the United States be launched on a space shuttle to an orbiting space station in 1976 to celebrate the 200th anniversary of the USA. Predictions were being made that, in 1986, a fleet of Mars ships would lead the first manned expedition at a cost of $75 billion (1969 rates, that is).

As the Apollo 11 trio embarked on their staggering 38-day, 25-country world tour, they were probably wishing they could go back to the Moon. Also wishing to go to the Moon were Apollo 12's Pete Conrad, Richard Gordon (or Condom as one newspaper typo described him) and Alan Bean. Three Navy pilots and good friends all. Having taken the trouble to land on the Moon, the crew would crash the lunar module onto the Moon's surface, to create a moonquake which could be measured by a seismometer deployed by them. "Apollo's smash hit"[1]; "America plans quake on the Moon"[2].

Meanwhile, NASA's chief geologist, Eugene Shoemaker, was resigning because he felt too much emphasis was being placed on getting men to the Moon rather than on the exploration itself. Science astronauts were resigning, too. "There's not enough science in the programme," one said. Finally, the Soviets were staging a remarkable troika Soyuz mission, which didn't seem to get anywhere, except to mark the first time that seven people were in space at one time. One cartoon summed up the feeling of aimlessness: "How do you know our mission was successfully completed if we don't know what we were supposed to be doing," asked one cosmonaut of another[3].

The Apollo 12 crew knew what to do: make a pinpoint landing in the Ocean of Storms, just yards from the unmanned soft lander Surveyor 3, which had made a bouncy landing in April 1967. Surveyor 3 also became the first moon-scooper. The Apollo 12 mission, with its nautical flavour – flight badge showing an ocean clipper sailing to the Moon and the mission names of *Yankee Clipper* and *Intrepid* for the command and lunar modules respectively – was a victim of the success of Apollo 11. "Man's second lunar expedition seems almost routine"[4]; "A scientific business trip to the Moon," said NASA. Budget cuts meant that the countdown was started early to save cash! The mighty budget axe weilder himself, however, would be at the space centre to bid Apollo 12 farewell. NASA was keen to have President Nixon there, it needed as much support as it could get. It also needed an on-time, flawless lift off to impress the President. NASA made sure it got the on time bit right but at one heck of a risk to the crew and mission.

Having raced against the clock to repair a leaking hydrogen tank in the service module's fuel cell system, NASA started the final count on 14 November 1969. At 11am, the weather was appalling with torrential rain and a thunderous black sky. President Nixon was shivering, ruffled up in his light mack. You could hardly see the launch pad. Yet the count continued. A cataclysmic blast-off followed and the mighty Saturn disappeared almost as soon as it had cleared the tower.

A drenched President Nixon (centre) waits for Apollo 12 to blast off. Astronaut Frank Borman seems to be praying for a good launch, knowing it ought to be scrubbed. NASA administrator Thomas Paine is next to him.

Conrad, in happy, gravelly and clipped tones, announced the roll programme. At T+36 secs, observers saw a bolt of lightning strike the launch pad. Almost immediately, the electrical systems on board *Yankee Clipper* went haywire and crackling static was heard over the voice link. Conrad matter of factly announced, "OK we just lost platform gang. I don't know what happened here. We had everything in the world drop out . . . fuel cells, lights and AC bus overload 1 and 2, main bus A and B. . . ." The test pilot Conrad was relaying as much data as possible before he bought the farm, according to an observer reported in *Aviation Week and Space Technology* magazine. Bean worked feverishly to restore systems as almost every light came on on the control panel. "It looked like a Christmas tree," said one of the crew. Calm returned as the Saturn powered its way ever upwards. Conrad remarked, "I'm not sure we didn't get hit by lightning". Hit by lightning? Nonsense said NASA, now facing a storm of its own, as everyone criticised the decision, seemingly made because Nixon hadn't to be disappointed. Apollo 12 had come close to a $300 million abort in the Ocean.

Soon, Conrad was calming things down, while Gordon chipped in, "I think we need to do a little more all-weather testing". Cap com astronaut Gerry Carr in mission control replied, "Amen to that. We've had a couple of cardiac arrests down here too". Apollo 12 replied, "We didn't have time for that up here". After the first stage fell away, Conrad amusingly described the SII stage as "chugging along minding its business". The drama was over in near space but the space in the newspapers was full of it, the media was having a field day.

"Moon Storm"[5]; "Off in a Flash"[6]; "Lunarnauts Ride the Lightning"[7]; "Apollo Crew Escape Blast off Disaster"[8]; "Blackout"[9]; "There's an awfully pretty Intrepid out there; let's go get it," said Conrad as the transposition and docking manoeuvre commenced.

The ride to the Moon was easy going and rather lighthearted, in stark contrast to the heaviness of Apollo 11. Conrad played his country and western tapes. "That wasn't half bad", said a cap com, "It was all bad". The astronauts slept for ten hours during their first 'night'. Later Yankee Clipper went off course. Deliberately. Previous Apollo crews headed for the Moon, in the knowledge that if something bad happened, they were on a free-return trajectory, one that would bring their spacecraft back home, without any engine firings. Because of the location of the landing site, Apollo 12 would have to place itself under greater risk by entering a hybrid trajectory that did not guarantee a free return.

Conrad was enjoying some of the home comforts of an Apollo, compared with Gemini, like shaving, having hot water and being able to brush his teeth. "All dressed up and nowhere to go," said Houston. "Oh," replied Conrad, "we are going someplace, and it's getting bigger all the time." The crew transmitted friendly TV shows for viewers, providing a tour of the two spacecraft. Mission control passed on family news and world news, choosing to ignore any references to the Vietnam War and protests against it. Meanwhile, what was assumed to be a discarded payload shroud of the S4B stage seemed to be trailing Apollo 12. Houston said that the unidenitified object was the back-up crew racing them to the Moon.

Apollo 12 heads for the thunderstorm.

Apollo 12 disappeared round the far side of the Moon and emerged from the other side minutes later, with Conrad announcing in his usual manner that they were in orbit, "Yankee Clipper with Intrepid in tow has arrived on time". Later he added, "like everyone else who has just arrived, all three of us are plastered to the windows". The commander was having problems, however. Biomedical sensors which were

stuck to his body were causing sores and itches that were "driving me buggy," he said. Soon, Conrad and Bean were inside *Intrepid* and in the midst of the PDI burn. Landing was due at 7.54 am GMT.

At the time I was sitting in car parked in a road in Chelsea waiting for a pub to open. The pub was to be a location of the TV interview to be linked with the first Apollo 12 moonwalk, with people who had Pan Am Tickets to the Moon. I was one.

Listening to the ebullient chatter coming from the the dynamic duo as *Intrepid* made its way down, it was easy to mistake the high-spirited shouts, coming from my small crackling transistor, for cries of disaster. Conrad, peering out of the window, looking for the guiding landmarks to steer *Intrepid* down to within walking distance of the Surveyor, shouted, "I think I see my crater. I'm not sure. . . . There it is! There it is! Right down the middle of the road! . . . Amazing . . . Fantastic." Bean cried back, "You've got loads of gas, come on down Pete . . . He's got it made . . . Come on in there . . . Contact, light!" Squeals of delight were heard from *Intrepid.*

Intrepid heads for its target.

There was no formal "Storm Base here, the Intrepid has landed", just "Holy cow, it's beautiful". With a crew like this, the moonwalks are going to be a wow, thought observers, especially with the first colour TV camera on the Moon.

The moonwalk started well, with the diminutive Conrad, at 5ft 6ins tall, parodying Neil Armstrong's words. "Whoopee . . . man,

Bean carries some equipment. Or is it Conrad?

that may have been a small step for Neil but it's a big one for me!" he shouted, as he dropped onto Moon, missing the footpad completely, creating a small cloud of dust. His faceplate was seen in colour by the TV camera. "One of the first things I see, by golly, is little glass beads", he said. A strange humming noise was heard. Conrad was humming, then singing to himself, much to the delight of the TV audience. He giggled and chuckled as he almost slipped over small potholes in the very dusty surface. "I'll tell you one thing, we're going to be a couple of dirty buggers". It was time for Bean to come down. "Don't lock it," said Conrad as his pilot closed the hatch. Bean dropped onto the surface and Conrad said, "Welcome aboard. . . . hustle boy, hustle, we've got a lot of work to do."

Bean's first job was to move the TV camera to a tripod away from the lunar module. He thought it would be a good idea to pan it around the surface and pointed it towards the Earth, for viewers to see themselves. But the Earth was too near the Sun in the sky and the inevitable happened. A public relations disaster at the moment of triumph. "The Moon Walk Blacked Out"[10]; "It's a Moon Black Out"[11]. Bean was to tell me later that they didn't honestly think that the sun would hurt the camera. He spent 30 minutes trying to rectify matters but had to abandon the attempt, only after hitting it with a hammer! "Guess what I see sitting on the side of the crater. The old Surveyor. . . . Good old Surveyor. It can't be more than 600 feet away," said Conrad. It was actually 700 feet away. Intrepid had indeed made a pinpoint touchdown. But with nothing to see, the audiences switched off.

"The Conrad and Bean Laugh In"[12] was the headline given by one paper parodying a cult US TV show, 'Rowan and Martin's Laugh In'. "Moonshow Stars Keep Walking – and Talking"[13] noted another. They had to. For the benefit of mission control and dwindling TV audiences, the third and fourth men on the moon had to keep a continuous commentary going so that their movements could be traced. Apollo 12 didn't quite capture the imagination. One cartoon showed a father peering blearly from his bedclothes as his son asks at the bedside, "Daddy, it's two o'clock. . . . what do you think the astronauts are doing now?"[14]. So, the second walk was rather an anticlimax, despite the high jinx, rendezvous with Surveyor 3 and the astronauts' footprints being recorded by the seismometer they had just deployed on the surface.

The Surveyor was a brownish colour but a mirror seemed unscathed after two and a half years in the harsh lunar environment. Bean cut off bits of the unmanned lander for examination back on Earth. The astronauts even rolled moonrocks down the sides of craters, like kids with snowballs. They jumped around like slow motion giraffes, they said. It would have made good TV. Pity. "The Billy Bunter style jokiness of the lunarnauts has led some to think that moon dust may conceal reserves of liquor," said one unamused paper[15]. We were to learn soon afterwards, too, that Conrad had once fallen over. This could have had disastrous results if his suit had torn. But from the antics of later moonwalkers you would be forgiven in wondering what the fuss was about. The astronauts looked as though they were coated in coal dust as they prepared to enter *Intrepid*, only after a suggestion by Conrad that they stay longer because their oxygen reserves were still good.

As usual, the lift-off caused a storm of its own on the Ocean of Storms, probably knocking over the US flag and swirling dust over $20 million worth of equipment and rolls of colour film the astronauts forgot to bring inside. Once they had realized, it was too late because their portable life support systems were outside too! "Away we go . . . looking good . . . we're right down the pike . . . nice ride . . . thrust is right in there . . . wow, this is a hot machine".

The rendezvous was perfect and made spectacular viewing, thanks to a camera on *Yankee Clipper*. Viewers could see the faces of *Intrepid*'s crew and the thrusters firing as the lunar module jockeyed its way towards the waiting Dick Gordon, who was to perform the docking. "Stand by to receive skipper's gig," said Conrad. "Aye, Aye, sir," replied Gordon. Conrad stared at the service module's S band antenna and didn't amuse his bosses by declaring, "Do you suppose that's where it got hit by lightning? Look at the top of it. It's burned". Bean rubbed it in, "Yep, I believe that's where it got hit". Millions of TV viewers 'docked' with *Intrepid*. "Contact," said Conrad," Super job Richard".

Gordon became so concerned about the moondust-ridden objects that were entering his command module, that Conrad and Bean decided to perform the first space streak,

Walking at the Ocean of Storms.

Working at the Ocean of Storms.

before it became a fashion on Earth. They removed their spacesuits and floated into *Yankee Clipper* wearing nothing but their communications headsets. Next, the ascent stage of *Intrepid* was discarded, fired its engine and committed suicide, crashing into the moon as planned at a speed of 3,500mph. The resulting moonquake was recorded by the Apollo 12 siesmometer and continued for over 30 minutes both mystifying and exciting scientists. It was as though, said one scientist, *as if* one had struck a bell once and heard it reverberating for 30 minutes. "Is this God's Gong?"[16] asked one newspaper idiotically.

Yankee Clipper leapt out of orbit and sailed Earthwards. Almost immediately, probably being aware of the effect of the moon blackout, the astronauts broadcast a TV show that showed the receding Moon. Although the TV camera was a colour one, it didn't make any difference, said the astronauts because the moon was black and white, although it appeared to have hues of colour when they were on the surface. The return was smooth, lighthearted and jocular and even included an in-flight press conference.

Continuing the nautical flavour of the mission, Conrad sent a message to the commander of the prime recovery ship, *USS Hornet*: "Apollo 12 with three tall hookers expect recovery ship to make its PIM as we have energy for only one pass." As they approached the Earth, their home planet eclipsed the sun and resulted in what the astronauts decribed as the most spectacular sight of the whole mission. The entire thin film of Earth's atmosphere was illuminated and flashes of lightning could be seen on the darkened side. The command module plunged into the upper atmosphere and before the re-entry communications blackout, Bean said, "Stop the world, I wan't to get off Apollo". Minutes later he received a cut over his eye when an unrestrained camera landed on his face. A pair of scissors, too, just missed him. But Apollo 12 didn't miss by much and was just 15 seconds late.

Reaction to the flight was not surprising and the *Daily Mail* summed it up nicely: "It wasn't any easier to land on the Moon for the second time than for the first but it seemed an awful lot more ordinary. Less worth losing sleep over. And in months to come, while Armstrong and Aldrin have their names safely in the history books, we shall be saying who were those fellows who got there second?"[17].

The astronauts had to wear the BIGs and enter a special container. They were shipped to Houston's Lunar Recieving Laboratory from where they were released to the outside world twenty days after splashdown. All important film take during the journey had been whisked to the laboratories soon after splashdown and after the now-traditional few days wait, the mission photos were released. One summed up the mission well – it was an astronaut inspecting Surveyor with *Intrepid* in the background on a hill not far away. But was the astronaut Bean or Conrad? It was no good asking the moon men themselves because often they were unable to identify who was who in the pictures. Because of this, it was decided that on future moonwalks, the commander would wear distinctive red stripes on his helmet and the legs and arms of his spacesuit. These were called, disdainfully the 'PR stripes' by the Right Stuff astronauts. In fact they were extremely useful and NASA Public Affairs man Brian Duff calls them 'his' contribution to Project Apollo. The first striped astronaut was James Lovell, commander of Apollo 13.

Apollo 13

38th manned spaceflight
11 April 1970
5 day 22 hr 54 min 41 sec

The Main B Bus Undervolt

WHEN I ARRIVED AT COCOA AND retrieved my rucksack from the luggage hold of a Greyhound bus, to walk across the causeway to Cocoa Beach, I had little doubt that I would get to see Apollo 13 lift off on time during a brief visit to the Kennedy Space Centre. So did the thousands of other people around. With just two landings under its belt, America had come to expect on-time lift-offs and milk runs to the Moon. They got the on-time lift-off and a milk run – until, that is, T+55 hours 55 mins 20 secs. Just how fate and luck, call it what you will, played such an important part in Apollo, is illustrated by the fact that the Apollo 13 explosion should have occured on Apollo 10, which was scheduled originally to have carried the errant oxygen tank, and would have put paid to any moonlanding by Apollo 11. Also Alan Shepard, not James Lovell, would have commanded Apollo 13 had not officials felt that Shepard needed more LM training.

Despite the almost brash confidence pervading the space centre, there was a little pre-launch drama when 'lucky' command module pilot Ken Mattingly was dropped from the flight two days before lift-off. He had been exposed to German measles by back-up lunar module pilot Charlie Duke and doctors feared that if he caught the mild illness in flight, it could impair his faculties. Working round the clock in the Apollo simulator, back-up command module pilot Jack Swigert was put through his paces and passed fit to fly. Mattingly was angry and upset; probably until T+55 hrs. So, emerging from the Manned Spaceflight Operations Building on 11 April 1970 at a reasonably comfortable time in the morning were the spacesuited James Lovell, commander, Jack Swigert, command module pilot – who had forgot to sign his tax papers – and the slightly oriental looking Fred Haise, who stopped briefly to look at the crowd of wellwishers before being the last to enter the transfer van with its Apollo 13 emblem emblazoned on the doors.

The press site was pretty busy, not quite as for the flights before, but busy enough. The BBC's James Burke had at last been let out of the studio to see an Apollo take-off. British vets Reg Turnill of BBC TV news, Angus McPherson of the *Daily Mail* and the *Mirror*'s Ronnie Bedford sat at their favourite positions in the press grandstand.

The cloud was high and a bit murky as the afternoon wore on and the tumultuous lift-off approached at about 2.13pm. Except I don't remember seeing it. Normally, from the press site about three miles from the pad, a Saturn looks about as high as a small match held at arms length. Using a 35mm camera with an ordinary lens, I became too snap happy and watched the entire show through the lens, which made the small rocket even smaller! The indescribable noise and small

Fred Haise enters the transfer van. Deke Slayton is on the left. (Author)

earthquake were something else, especially as they arrived when the Saturn was a good 500 feet up, having, seemingly, left the pad in serene silence. Despite the loud speaker announcements, the drama of a premature second stage engine shutdown passed most people by.

Parking orbit, translunar injection, transposition and docking manoeuvre all seemed so simple. Even the TV networks were not taking the televised shows from Apollo 13's command module *Odyssey*. The last of these was transmitted at about T+55 hours. Lovell said goodnight with a "this is the crew of Apollo 13 wishing everyone down there a nice evening". He could have been saying goodbye for under *Odyssey* was a time bomb and the time was running out. Inside the fourth bay of the service module was an oxygen tank, part of the fuel cell system, providing electricity. This had been fitted to Apollo 10's service module but had been removed after experiencing some trouble – it had even been dropped on the floor! It was then installed in Apollo 13 and during a pre-launch test in March, excessive electrical loads welded heater switches shut when liquid oxygen was being boiled off in an effort to empty the tank. The heater switches were not checked. The next time the lox tank was filled it was a bomb. And it exploded 205,000 miles away from Earth.

Each crewman heard a thud and felt some vibration – Fred Haise was checking out the attached lunar module *Aquarius*. Jack Swigert, lying in the centre seat of *Odyssey*, saw his panel light up with urgent messages of doom. He said calmly and somewhat incongruously, "Houston, we've had a problem." Houston didn't catch any urgency in the voice because they wasn't any. "Say again please," said cap com Jack Lousma. By this time Lovell had floated onto the scene and a brief look told him he wasn't going to be the fifth man on the Moon. His resigned disappointment could be detected when he replied, "Houston, we've had a problem, we've had a main B bus undervolt." The astronauts did not need to know that half the service module had been blown apart to know that they were in serious trouble. No electricity, no water, lack of life support systems. If that had happened after the lunar module had been jettisoned, when *Odyssey* was on the way home, the crew would have died. They still had *Aquarius* and the ensuing drama of Apollo 13 that gripped the world for three days unfolded.

Soon after the 'thud', as the astronauts described it, Lovell reported that he could see what he thought was a gas venting from the service module. Houston tried to reassure the crew, "OK 13, we've got lots of people working on this, we'll get some dope as soon as as we can have it. and you'll be the first to know." Within an hour cabin oxygen pressure

The explosion ripped out part of the service module but the damage was only visible towards the end of the flight when the module was discarded.

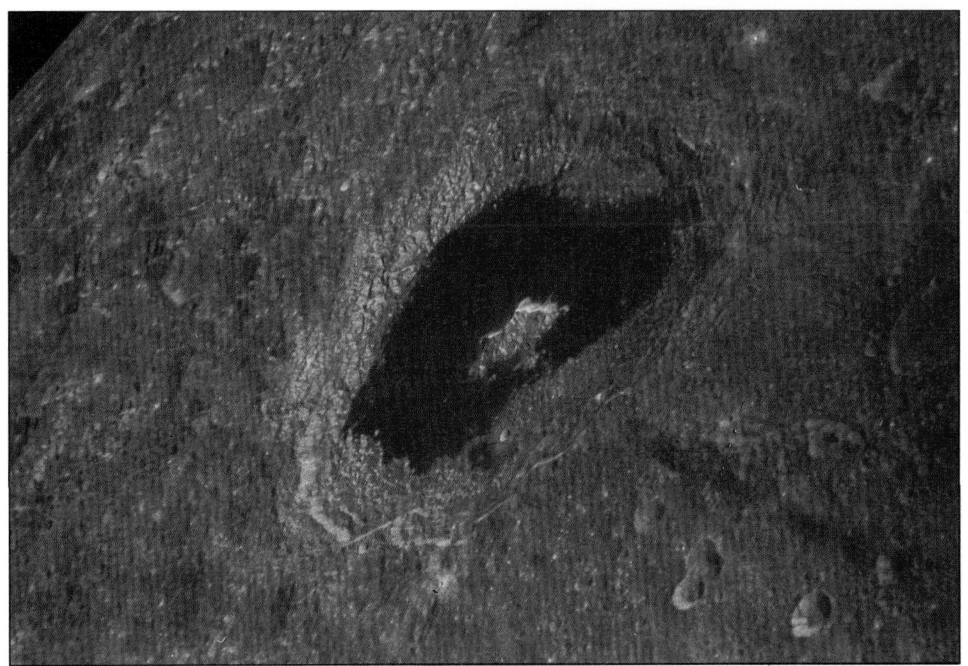

Haise and Swigert get their first and only views of the Moon, the farside and the crater Tsiolkovsky.

was dwindling to nothing. Houston reported, "Its slowly going down to zero, we're starting to think about LM lifeboat". A voice from the spacecraft replied that it was something that had not gone unmentioned. The crew was ordered into the LM before the power was completely gone in the command module. They had about 15 minutes in which to power up *Aquarius*.

It was only shortly after this that NASA announced the obvious officially, that the moon landing was off. Briefly, *Aquarius* was to use its descent engine to perform mid-course manoeuvres to ensure a safe and reasonably fast passage home. It wasn't a simple case of turning round; the crew had to go all the way to the Moon, once around it and then back to Earth. *Aquarius* would be used to provide life support, in conjuction with much jury rigging from the crew. The conditions were not good. It got cold and wet inside the command module. And the crew had only hours to spare by the time they got home. *Aquarius* fired its descent engine for about 30 seconds to bring Apollo 13 back on a free return trajectory, one which at least guaranteed that the command module would hit the Earth – somewhere. Apollo went around the Moon and the rookies, Swigert and Haise, squashed their noses against the windows to get a brief peek of it, taking some pictures at the same time. During the brief pass, the three became the world's deepest space travellers, journeying 248,655

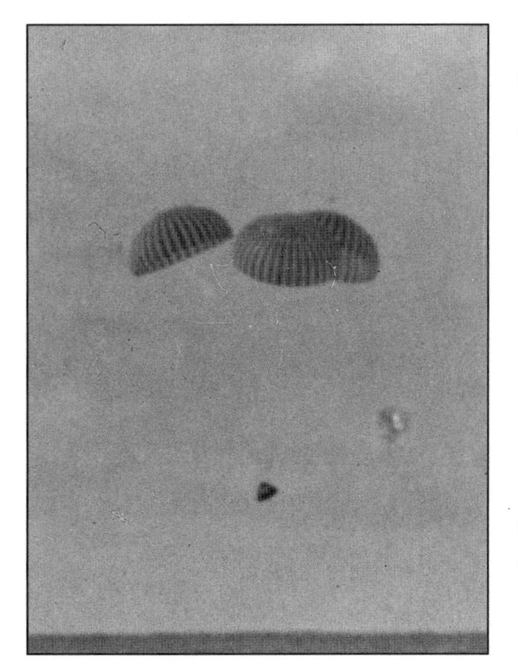

Relief as Apollo 13 returns safely.

President Nixon is on hand to honour the brave crewmen.

miles from Earth. A second burn followed, to speed the return and aim the 'lifeboat' towards the Pacific – somewhere.

Then new trouble hit. Lovell reported that carbon dioxide levels were getting high. Without some solution, the crew would gradually go to sleep and die. The item that illustrated the jury-rigging effort and resourcefulness that went to help Apollo 13 survive was the new air conditioning unit assembled using bits of spacesuit hose, lunar module components and the lithium hydroxide canisters from the original system, which was not being allowed to do its job properly because of the crisis. The command module was getting cold and the crew was moving about using a flashlight. The crew had a great difficulty sleeping and were taking dextrose tablets to keep reasonably alert. It was then discovered that Apollo was aiming for too shallow an approach into the Earth's atmosphere, with a danger of bouncing off into oblivion. A third and final LM burn fine-tuned the path to allow a slightly steeper approach.

The mood of the crew was surprisingly good considering their ordeal. Only now and again did something slip out. Before the third burn, Lovell, said, "I hope the guys in the back room thought this out right". Reacting to requests to give latest readings from the spacecraft, Lovell said sharply, "We've got to establish a work/rest cycle here. We can't just wait around here and just read figures all the time, we've got to get some sleep, so take that into consideration." The conditions in the spacecraft were now pitiful. It was near freezing. Water on the windows was turning to ice. Fortunately the ordeal was nearly at an end. Small thrusters fine tuned the re-entry path. *Odyssey* jettisoned the service module and it was only then that the crew saw the devastation. "It's incredible," said Lovell. "There's one whole side missing". Mission control replied with the somewhat dubious choice of response. "Man that's fantastic!" A push-pull technique deployed *Aquarius*. "Farewell, Aquarius and thank you," they said, "She was a good ship". Now on what was left of battery power, the lone command module hurtled towards the atmosphere. Radio contact was lost, as usual as re-entry commenced, and the world prayed while James Burke sat in the BBC studio back in London, eyes closed and every finger crossed. Then through the crackle came the simple remark, "OK Joe," addressed to cap com Joe Kerwin. *Odyssey* splashed down only about three miles from the USS *Iwo Jima* recovery ship. Once on board, the crew bowed their heads, while the ship's chaplain read a prayer. The world breathed a sigh of relief. It was over.

Although the astronauts had been aware of the danger that they had been in, when they returned, they were frankly surprised by the intense media coverage of their plight.

"Fight to Save Moon Men"[1]. Accompanying a cartoon showing hands held in prayer with the Moon in the sky, a newspaper proclaimed, "The Hours of Hope"[2]; "The Odds are Against Them – But NASA Hush-up is denied"[3]; "The Grim Battle"[4]. As the crew approached the Earth in their crippled command module, it was "The World Waits"[5]. The day after splashdown there were joyous headlines including, "Beautiful!"[6].

President Nixon called a national day of Thanksgiving and in Churchillian tones, declared that in the Apollo 13 drama, "Never have so few owed so much to so many," alluding to the thousands of engineers and contractor staff who had worked 24 hours a day designing trouble shooting procedures to keep the crew alive. The irony was that Nixon had only recently been responsible, once more, for cutting the space budget, one result of which was the issue of over 5000 pink redundancy slips to Kennedy Space Centre staff just days before Apollo 13 took off.

Apollo 14

40th manned spaceflight
31 January 1971
9 day 0 hr 1 min 57 sec
3 hr 31 min on Moon
98 lb of Lunar samples

Rookies
head for Fra Mauro

AT THE TIME OF THE APOLLO 10 MISSION in May 1969, the back-up crew of Gordon Cooper, Donn Eisele and Edgar Mitchell, looked an odds-on bet to take Apollo 13 to the Moon, to be followed by James Lovell, Fred Haise and a command module pilot from the Apollo 11 back-up crew who was replacing the retiring William Anders. This might well have happened had it not been for the fact that America's first man in space, Alan Shepard, had forced himself back into the reckoning, after being grounded by an ear disease that effected his balance and stopped him from flying airplanes, let alone spacecraft. Cooper and Eisele did not help their causes either. Cooper liked to race fast cars and Eisele was the first astronaut to divorce publicly, a very un-*Life Magazine* thing to do, although there was a certain disenchantment with NASA anyhow. Shepard took over from Cooper and Stuart Roosa from Eisele and both joined Mitchell for Apollo 13, rookies all, in terms of orbital flight, with only Shepard having made a spaceflight of sorts, experiencing five minutes of zero G. Yet, they were flying to the Moon. It was then decided that Shepard needed more training on the lunar module and the Apollo 13 and 14 crews, the latter heading for Taurus Littrow, were switched.

After James Lovell had returned his crew from disaster after drawing the short straw, Shepard's Apollo 14 mission, diverted to Fra Mauro, was delayed from 1970 to January 1971, to give time for engineers to perform $12 million worth of modifications to the command module. As the crew went into intensive training, which included a tour of a 15 million-year-old West German meteorite crater called Noerdlinger, resembling Cone Crater in the Fra Mauro hills of the Moon, NASA chief Thomas Paine resigned and budget cuts were made. This meant that Apollo would stop at 17. It also meant that 3,000 NASA staff would be laid off. Adding fuel to the fire, on 24 September 1970, was the return to Earth of a capsule from the Soviet Luna 16. It had landed on the Moon, scooped up some soil and now that soil was back on Earth. Manned flights to the Moon were unnecessary, said the Russians. Luna 17 followed and carried with it a real Heath Robinson affair, the Lunakhod lunar rover.

As Apollo 14's day approached there were grim reminders of the Apollo 1 fire. Ed White's widow announced her engagement and Grissom's sued Apollo spacecraft builders Rockwell for $4 million. Apollo 14 back-up commander, Eugene Cernan, meanwhile, was being fished out of Banana River near the Kennedy Space Centre after crashing his helicopter. Even with four manned Moon landings in the offing, the area around the Space Centre was becoming like a ghost town. "As the Moonport faces closedown, only Disney and a defrocked minster may be

Launch brightens the teatime gloom.

left",[1] wrote Ross Mark of the *Daily Express*. The party was breaking up. Adding to the gloomy atmosphere was the 'Poor People's March', that was descending on the Cape. What publicity about the flight there was, inevitably centred on Shepard, at 47, the oldest man in space. "Al, the man who wished for the Moon . . . and got it!"[2].

There were times during the Apollo 14 mission when it looked as if Al wouldn't be getting the Moon. They started during the countdown. Three o'clock in the afternoon at the Kennedy Space Centre was dark and gloomy. As stormclouds billowed around the sombre scene, they evoked memories of Apollo 12's eventful ascent. The countdown was stopped and was held for 40 minutes, which was peanuts to Shepard who had endured a 4hr 22min wait for his first flight. "We're in fine shape up here," he said. The launch was cleared but there were some who questioned whether it was worth the risk. It was to NASA because a cancellation would have meant a delay to March, to meet the right lighting conditions at Fra Mauro. "Thank you very much. We'll give it a good ride," said Shepard as he prepared to lift off. He sounded very calm. How calm we would not know, because his heart monitoring sensor had come loose and he became one of the few astronauts whose heart rate at lift-off was not widely publicized.

To the half a million or so spectators at the space centre, the thrilling lift-off lasted less than a minute before the thunderous rocket was obliterated by clouds. Among the observers were Roosa's mother, who had bad him farewell as he left for the pad. To Roosa's wife the launch was, "like being pregnant. The closer you get to the crisis the more uncomfortable it gets". The irony of Mrs Mitchell's remarks would only become apparent long after the flight. "Separation from your husband is difficult but makes you keenly aware of the beauty of every moment you are together."

"It's great. We're having a ball," said Shepard before TLI (trans-lunar insertion). After TLI and transposition and docking manoeuvre, Roosa cast a pall of gloom over the mission. "We are unable to capture," he said. "2am. Trouble for Apollo"[3]; "2am. Moon Crew Hits Trouble. If this is a flop, it could be the end"[4]; The astronauts were completely baffled. "I hit it pretty good but we didn't latch," said Roosa. It was beginning to to look like Apollo 14 was going to do a lunar loop,

Antares, *Roosa's elusive target, nestled inside the S4B stage.*

Mitchell heads towards Shepard's shadow.

as five desparate attempts failed. Finally, after 106 minutes, a sixth attempt was successfull. The crew and the watching world breathed a sigh of relief. Before lift-off, Roosa had attended a private Roman Catholic Mass. "Tell Stu, that session he had this morning paid off," said cap com Gene Cernan. "I believe it," replied Roosa. With command module *Kitty Hawk* now safely attached to lunar module *Antares* and *en route* for the Moon, the crew dismantled the docking probe and inspected it. "We cannot force it to malfunction at all," said Shepard. It was an unsatisfactory situation. It may have been better if the astronauts knew what the matter was, then they could have fixed it. As it was, extreme doubts were cast over *Kitty Hawk's* ability to dock safely with the *Antares* ascent stage after the lunar exploration. Would Shepard and Mitchell end up having to spacewalk over to *Kitty Hawk*, clutching their Moonrocks? The mission was close to aborting its prime objective. Peter Cole of the *Evening News* wrote: "Between now and Friday, mission control at Houston will make a decision on which the future of the American space programme depends ... if they decide to call off the descent of *Antares* into the Fra Mauro region of the Moon, they may be signing the death certificate of western man's exploration of space"[5].

There was already widespread public cynicism and apathy about Apollo. "When Americans are told daily about the diabolical state of their economy ... they look at NASA and wonder. They care more about spending money on cleaning the environment and the ghettoes ...," Cole continued. The environment in *Kitty Hawk* was at times puzzling. The astronauts said they were seeing mysterious flashes in their eyes. These were thought to be cosmic rays. Other than that, conversation was sparse to say the least. Shepard, Roosa and Mitchell were being dubbed, "The Silent Crew". The Earthbound Fred Haise called *Kitty Hawk*, "just to make sure you are all still around up there."

Later it was "Go For the Moon!"[6]; "Go ahead for the Moon!"[7]. The decision had been made. The silent crew became 'chatternauts' as they emerged from the far side of the Moon after the lunar orbit insertion burn. "Well, that turned out to be a piece of cake," enthused Roosa. "Gosh, it really is a quite a sight," said Shepard. "Wow. This is really a wild place up here." Mitchell remarked that the Moon, "looks like you could walk along and fall into nothing, just darkness". Roosa displayed some nice deadpan humour when he said, "We sure picked a clear day to arrive. There's not much haze in the air at all. You can see all the way to the horizon". Joining in the fun, Haise replied, "Incredible". All of it, said Mitchell, "just wets your appetite to get to Cone Crater". That Mitchell was so keen to get to this crater, explained his mood on the Moon later. He was lucky to get to the Moon, too. Early in the descent, *Antares* experienced serious difficulty locking its radar onto the surface. Alarms sounded and an abort was close. Mitchell desparately flicked a switch on and off. With just seconds remaining before an abort, the radar locked on. "Come on, come on.... Phew, that was close!" Shepard was as cool as a cucumber. You almost imagined him serving tea and sandwiches and asking Mitchell, "One lump or two?" Shepard had his critics and had had to ride above the 'old man' jokes. You had to hand it to him. As *Antares* came down, he said, "Starting down. We're in good shape troops". As the module settled on a slope at Fra Mauro, the ever so slightly excited Shepard said, "We're on the surface. OK, we made it. Oh, brother." His wife Louise remarked, "They can't call him old man Moses any more. He reached his promised land!"

They did call him an old man again. As Shepard slowly came down the ladder and dropped onto the footpad, cap com Bruce McCandless said, "Not bad for an old man". Shepard stepped onto the Moon and said, "It's been a long way but we're here", alluding more to the ten years between his spaceflights than to the distance travelled. After his arrival on the surface, Mitchell took a TV camera out of the side of *Antares* and placed it on a tripod yards away. TV viewers saw the whole lunar module on a gentle slope with a small hill in the background. Frustratingly, the astronauts rarely stayed in view for long. The pictures were somewhat grainy and lacking in resolution. To the astronauts, the scene was stark with the sun so low in the sky. A famous photograph later showed Shepard at the foot of the ladder one hand shading his eyes.

The astronauts got about the matter of fact business of deploying instruments which, in Mitchell's case meant detonating 'thumper' charges in the ground to provide work for the seismometer. The wheelbarrow also came out. This was a rickshaw-like two-wheeled trolley which was used to carry tools and samples. Many wags suggested that it was to carry Shepard when he got tired. A successful moonwalk over – despite Mitchell's slightly leaking spacesuit and lots of lunar dust which slowed them down and covered their spacesuits – the crew got some well earned rest. "Moonmen Go To Bed"[8]. One newspaper showed Mrs Mitchell 'touching' her husband on the TV screen: "My Man on the Moon"[9]. It was reported by the astronauts' doctor, Charles Berry, that Shepard's heart rate never rose above 113 beats per minute. "Most men get that excited parking their car," he said. His heart rate rose much higher the following day, however.

Tracks of Shepard's 'wheelbarrow'.

The next day's moonwalk was to be the highlight of Mitchell's life. A walk up the 300 foot slope of Cone Crater and down inside it, to pick up some interesting samples. After conducting some initial experiments and falling slightly behind schedule, the astronauts set off for the crater. "The next stop is the top of the crater," said the optimistic Shepard. Mission control said that they were worried about oxygen levels. The astronuats pressed on. But, like mountaineers in soft snow, they really struggled. Boulders made life difficult too, as the 'wheelbarrow' sometimes had to be carried. The astronauts lost their way. Looking at a map, Mitchell said "We're down here and we've got to go there." Gasping for breath and with his heart rate at 150 beats, Shepard suggested gently that they weren't going to make it to the top. Mitchell said firmly, "We can press on and make it". After plodding on a little further, mission control suggested that they turn back. A crestfallen Mitchell, urged, "Aw, gee whizz, let's give it a whirl. I think we'll find what we're looking for down there". Eventually, they were ordered to turn back. "I think you're finks," said Mitchell. Apart from being off course and in difficult terrain, the added problem was judging distances. The exasperated Mitchell later found that he had actually got to within yards of the rim of the crater. What little humour that the astronauts had displayed earlier, had evaporated.

But Al saved the day. "Golf on the Moon"[10]; "Puffed out Moon man plays golf!"[11]; After walking nearly a mile back to the lunar module and before re-entering it, Shepard stopped within view of the camera, seemingly to make a departure speech. He took an instrument from the 'wheelbarrow'

and pulled something out of his spacesuit pocket. "In my left hand I have a little white pellet that's familiar to millions of Americans," he said, dropping a golf ball into the dust. Then, using a makeshift club made from the handle of a tool and the six iron of a real golf club and using just one hand because his bulky spacesuit wouldn't allow him to use two, Shepard played a sand trap shot. "That looked like a slice to me, Al," said Fred Haise in Houston. He played another, shouting, "Straight as a die ... miles and miles and miles." In fact, the ball only travelled a few yards. Shepard tried one more shot before hurling a shovel, describing it as "the javelin throw of the century".

If there were any doubts about the integrity of the docking system on *Kitty Hawk*, they were dispelled when Roosa got a first time, hard dock with *Antares* just back from the surface. "Beautiful, beautiful," said Shepard. A successful service module engine burn followed and the astronauts were on their way home.

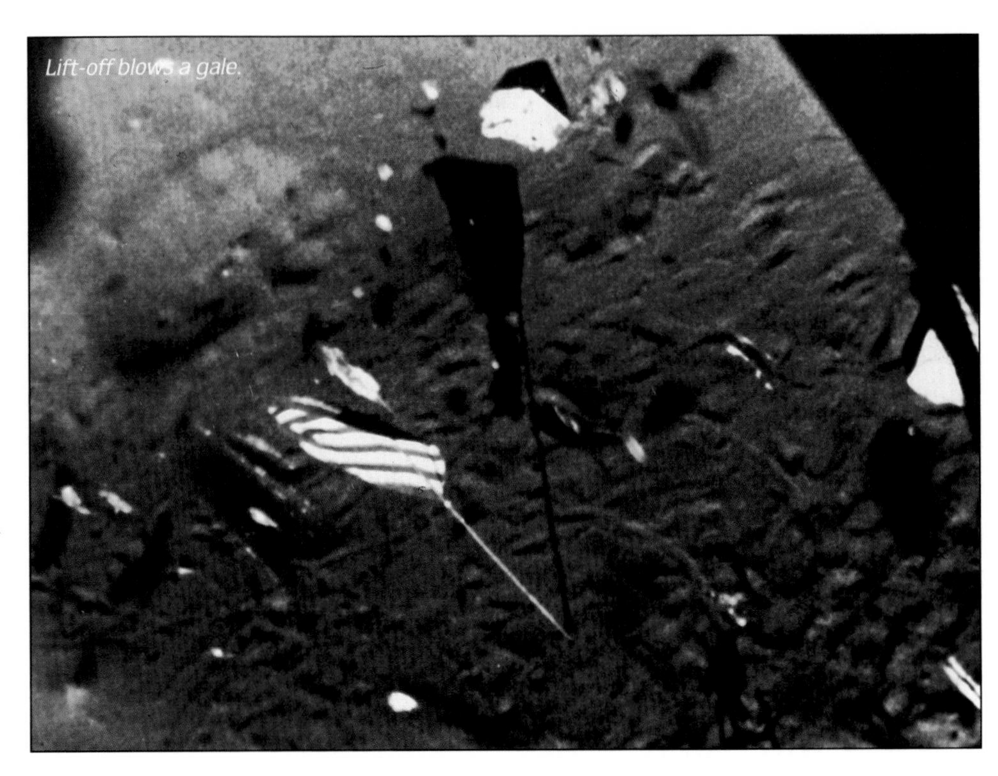
Lift-off blows a gale.

Mission control suggested that Shepard take some photos of the receding Moon with a large topographical camera. "Are you really serious? We're just ready to go to sleep here". Shepard and Mitchell then made the point, sleeping for a record 11 hours. In the 'morning' Mitchell told doctors, "We've had no medication, we are all in excellent shape, so just tell the surgeon to sit back in his chair and have a cup of coffee. We're fine". Later, the crew sent back a televised programme in which the usually taciturn Shepard made a plea for world peace. "It is our wish tonight that we can in some way contribute through our efforts in the space programme to promote a better understanding of peace throughout the world." In one of the few references, if not the first, to the Vietnam War uttered by an astronaut in flight, Shepard said, "We are reminded as we look at the shimmering cresent tonight, which is the Earth, that there is still fighting going on ... We are reminded that some of the men who have gone to Vietnam have not returned and are still being held there as prisoners of war". TV Viewers meanwhile watched 'The FBI', 'The Wonderful World of Walt Disney' and 'The Ed Sullivan Show.' No TV network took Shepard's show.

In fact, the whole flight seemed to be summed up by ITV's Alistair Burnet's question: "Well, was it all just a bore?" Milton Shulman in the *Evening Standard* wrote, "in the beginning there may have been heaven and Earth, followed by the Moon and stars, but nowadays they all take second place to the ratings"[12]. At least ITV's coverage was embellished by the ground-based Gordon Cooper – who should have been where Shepard was – whose Oklahoma drawl "incurred the potential wrath of Mrs Whitehouse" – has she really been at it for 20 years? – by describing, "ass ... ent engines". Cooper had also said on ITV that he considered Shepard unqualified to lead the Apollo 14 mission because of lack of experience and poor health. An Emwood cartoon showed Shepard being examined after the successfull splashdown by a doctor who tells the astronaut, "Heart rate OK, pulse OK, blood pressure great, but holy mackerel, your golf swing looks terrible"[13]. By the following April, Mitchell, too, was in the news again. "Moonman sued for divorce".

Apollo 15

43rd manned spaceflight
26 July 1971
12 day 7 hr 11 min 53 sec
66 hr 55 min on Moon
173 lb of Lunar samples

The Rovers of Hadley Base

THREE APOLLO MISSIONS WERE CANCELLED in 1970. They were identified specifically as Apollo 15, 19 and 20. This, theoretically, meant that Apollo 16 would become 15, 17 16, and 18 17. A mammoth crew reshuffle was in the offing and a desparate fight by those threatened to be dropped ensued, to get themselves retained on newly designated flights. Ultimately, sense prevailed and the next crews in line were kept in order. Commander David Scott and command module pilot Al Worden seemed safe. But the fate of Jim Irwin was in the balance. There was only one geologist in the NASA astronaut corps. His name was Jack Schmitt and he was down for Apollo 18, which was not to be made Apollo 17 after all. If anyone was going to the Moon, it was he. Someone had to give way. Some NASA executives felt it necessary to replace Irwin with Schmitt to ensure that the geologist explored the surface before the blade of the budget axe crashed down on Houston. Scott even went into training with Schmitt. Scott fought for Irwin's retention and was successful. It was left for Charles Duke and Joe Engle, the lunar module pilots of Apollos 16 and 17, to quake in their moonboots. One of them was going to miss the Moon. That Jim Irwin didn't miss the Moon was to have a dramatic effect on his life afterwards, perhaps in more ways than any moonwalker. It was also to have a dramatic effect on many people's lives all over the world.

Despite that fact the Scott and Irwin were heading for the most spectacular landing site yet, among the hills and valleys of Halley Rille, and were to drive a lunar roving vehicle, there was little build up to the mission in the media. The astronauts were preparing to make the most perilous journey in history, involving a steep descent over the mountains of the Moon, yet it all seemed so matter of fact. The dangers of travelling in space, let along to the Moon were, however, well illustrated a month before Apollo 15's lift off. "Disaster in Space"[1]; "Cosmonauts Perish in Space Mystery"[2]; "Man Takes One Step Too Far"[3]. Having spent 23 days aboard the Salyut 1 space station, three cosmonauts landed back on Earth. Rescue teams rushed to the capsule, opened the hatch and saw the crew lying "as if asleep". Had they been in weightlessness too long? Before re-entry, the flight cabin lost all pressure due to a faulty valve. But surely spacemen could survive this peril. Isn't that what spacesuits are for? The three men didn't wear spacesuits, just woolen tracksuits! Man had taken a step too far – from space safety.

Publicity about the safety of space travel then focused in a modest way on Apollo 15. "The target chosen for Apollo 15's landing on the Moon may just turn out to be impossibly hostile, littered with rocks as big as houses"[4]. Apollo 15's 'mission scientist', Joe Allen, himself an astronaut, with no hope of making it to the Moon, but who was responsible for lunar surface exploration operations said, somewhat euphamistically, "It's definitely worth the risk." Another risk was that Scott and Irwin might lose their way on the Moon. A drive in their rover one and a half miles from the lunar module would take them out of sight of it. A compass would be no use because the Moon has no magnetic field but, as ace correspondent Angus McPherson of the *Daily Mail*, reminded us, the rover had a robot brain to guide it and ensure that the astronauts knew where their Lunar Module was and that they didn't drive more than five miles away from it. Neil Armstrong walked just 300 feet from *Eagle* and here were astronauts three landings later, contemplating five miles. "Moon Mini Gets Set to Ride"[5]; "Lunar Buggy Ride to Solve Mystery of Mile Wide Rille. Longest Mission will cost $190 million"[6]; "Planning a Picnic for two – on the Moon"[7]. This last headline referred to the the rations fixed carefully inside the astronauts' helmets. Irwin got in a fast set of tennis before blast off. A set of tennis was to have a dramatic effect on Irwin after the flight.

The countdown and lift-off of Apollo was to have a dramatic effect on me, too. My build-up included attending astronaut parties at the Holiday Inn at Cocoa Beach, meeting people whose pictures used to hang on my bedroom wall. The night before lift-off, some press who wanted to face voracious mosquitoes in the sultry evening, were taken to within a mile of the pad, to take pictures of the Saturn as the service structure moved away, some 12 hours before launch. With the sun setting behind thunderous stormclouds directly behind the pad, the scene was breathtaking. The following morning, I was sitting close to the exit of the Manned Spacecraft Operations Building at the Kennedy Space Centre, camera in hand, like everybody else crammed into every nook and cranny around, waiting for the 'last photo opportunity'. The unspacesuited, grounded astronaut Deke Slayton came out to give the press the latest status reports and then the spacesuited astronauts, led by Scott, emerged into the glare of floodlights. Jim Irwin followed and Al Worden stopped to bid farewell to members of his a family. The first divorced man to fly into space, said goodbye to his two daughters. The astronauts waddled towards the transfer van which, as usual, had the mission emblem on its door. All the while, there was a constant clickety click and whirring of cameras. The van drove off and the press rushed for the buses for the drive to the press site. It was at least three hours before lift-off, so why the rush, I thought, nevertheless rushing for the first bus.

The sun rose to begin what was a typically sultry, perspiration-drenching Florida day. The white needle that was Apollo stood white and erect against a blue but misty sky. The atmosphere at the press site was like a carnival. Most people were wearing cardboard sunhats emblazoned with the name of Coca Cola and other drinks. Three men were about dissapear off the face of the Earth in a numbing blast off and it was as if we were watching Wimbledon. The real working press were beavering away in the press grandstand, while we more relaxed people could find a spot on the grass bank

Countdown . . . ignition . . . lift-off . . . tower

before it. I chose a spot on the bank of the barge-turning lagoon that leads to the Banana River. My friend, an ace British industrial photographer, Don Fraser, had lent me his expensive Hasselblad, which was safely secured on a tripod this time, to ensure I didn't watch the event through the camera as I had done for Apollo 13. I stood waiting and chattering among a group of extremely lighthearted people. The public affairs man droned on from the loudspeaker with vital information about the progress of the count. Finally, the moment arrived. "Twelve, eleven, ten." The sound of crickets erupted as the cameras went into action at the start of 'Ignition sequence start'. A small ball of rich orange flame appeared at the base of the rocket. Suddenly it exploded into a massive fire, spewing out clouds either side of the pad. It seemed to sit there for ever, churning away. "Oooo, oooo, aaahh, oooh ... whooohay," came the shouts. At last the Saturn was released. It rose slowly, majestically and in total, errie silence. There are those that say there isn't something Freudian in a lift-off – don't you believe it! Seven seconds after lift-off, the noise arrived. A rumbling. The sound of ten express trains rushing through a station. A shattering cacophony of whiplash cracks. It went on and on, as the rocket went higher and higher, trailing a tongue of flame and smoke twice its length, as clouds at least a thousand feet high billowed each side of the pad. The noise reached its crescendo at T+25 secs, began to diminish at T+40 and at T+60 secs was a gentle murmur in the sky. The sky was so clear, we could see staging. But soon the experience was over. The launch pad was empty and steaming from its drenching of cooling water. As people packed up their things, the awestruck feeling was astonishing. Nobody spoke for minutes. They just stared. But soon the laughs and giggles broke again and the carnival atmosphere returned. But for Scott, Worden and Irwin the experience was beginning.

clear . . . anticlimax. (Author)

By the time command module *Endeavour* approached lunar module *Falcon* for the transposition and docking, the crew had already gone into the history books "as the least vocal crew ever to journey into space". No words were uttered during TPD for fully five minutes until Worden reported, "We have a hard dock". Then came the first of several niggling problems, this one seemingly threatening the landing. A short circuit had resulted in signals in the command module indicating that the service propulsion system was misbehaving. "Apollo Hitch after Perfect Blast Off . . . After the most spectacular and trouble free launching ever seen at Cape Kennedy. . . ."[8]. The hitch turned out to be a false alarm, according to the press. This seemed premature because a short cicruit is a short circuit. Before the moon landing mission could be given the go ahead, Scott had to fire the SPS engine for one second, to ensure that it would work. It did, because Scott reported "we have four point seven, five point three". Of course it had worked. "Mini Blast Sorts Out Space Fault"[9]. "We'll proceed with a nominal mission," said Joe Allen. The next little problem came when Scott and Irwin entered *Falcon* for the first time, to find pieces of broken glass floating all over the place. The outer glass shield of the radar altimeter indicator had shattered mysteriously. The vacuum cleaner came out. The main worry was that some unfound glass could have floated into the docking system, to foul things up later. Things were getting a little uptight aboard *Endeavour*. Just before going to 'bed' Scott reported that they was a sudden deluge of water in the command module. The cap com quietly and casually told Scott that Houston thought that a fitting was lose on the water tap and there was a way to tighten it up. Scott snapped back sharply, "OK give it – quick!" The "flood" seemed to have been the last straw. "All uptight in space!"[10] reported one paper. *Endeavour* sailed into orbit and Scott cheered up. "Endeavour is on station and what a fantastic sight," he said.

Scott soon became uptight again. *Falcon* and *Endeavour* failed to separate as planned behind the Moon. An umbilical connector had come loose. Falcon separated at the second attempt and later began its steep 25 degree descent over the 13,000 foot high mountains, armed with enough extra fuel to allow an additional 17 seconds of hover over Hadley.

The arrival was absolutely flawless. "Almost like a simulation." said Houston. Scott's formal and utterly calm announcement after touchdown was "OK Houston, Falcon is on the plain at Hadley". The 11.16 had arrived on platform four. Yet, *Falcon* had come in a little high and for the last 50 seconds, Scott was 'blinded' by clouds of dust. He joked about the terrain later, "It's very hummocky. In this kind of terrain, you can hardly see over your eyebrows". Scott saw over the top of *Falcon* later, when he made a unique 'stand-up EVA', poking his head out of the top of the docking port. He then gave a 360 degree description. The attention turned to the first moonwalk, to take place on Saturday afternoon in Britain. "Motorists on the Moon. . . . On a day Britons traditionally travel to their cars and head to the seaside in a traffic jam called contentment, the astronauts take man's first drive on the Moon," wrote Ross Mark in the *Daily Express*"[11].

David Scott set foot on the Moon and, at last, a moonlander spoke like an real explorer."As I step out here in the wonder of the unknown at Hadley Plain, I realise that there is a fundamental truth and law of nature. Man must explore and this is exploration at its greatest". Irwin came down the ladder and said, "Oh boy! It's beautiful out here!" The TV coverage was spectacular. Viewers could see the smooth but high mountains and gently rolling valleys. The BBC did not mount an Apollo Special to cover the moonwalk but included coverage of it on the sports programme 'Grandstand'. Once the strangely fragile Lunar Rover had been deployed, the astronauts took viewers for a ride and just to ensure no one complained about the moonwalk being included on a sports programme, one of them said, "Man, oh man, what a Grand Prix this is!"

The moonwalk clearly exhilarated the astronauts, whose almost continuous dialogue was as breathtaking as the TV views. This was exploration at its greatest. The six hour moonwalk involved drives to Elbow Crater, the St George foothills and to the Hadley Rille itself. The Lunar Rover gave the astronauts a 'rock 'n roll' ride. The terrain was peppered with hurdles. "I see I'll have to keep my eyes on the road," said Scott. An electrical problem meant that the astronauts lost front wheel drive, which made for some hairy moments, particularly on the dusty slopes. "Wow, we're bucking like a bronco! It

Irwin salutes the flag.

Irwin loads the Rover.

feels like we really need these seatbelts". Each time the rover stopped, a remote TV camera panned around the moonscape, showing what the astronauts were up to. It was controlled from the ground by an engineer at Houston. "Oh, there's some beautiful geology out here," said Scott, describing the Halley Rille and the tracks that had been made by two rocks that had rolled down its steep slopes. Irwin described Elbow Crater as "unreal, the most beautiful thing I ever saw". One TV view reproduced in the newspapers the following day, showed Irwin helping Scott to his feet and it looked just like the negative of a photograph of Scott of the Antarctic. The exploration did not just involve lunar drives but a lot of hard work close to *Falcon*. This included drilling into the surface, which Scott found extremely difficult. "The base at Hadley is firm," he said. The astronauts were getting very tired towards the end of the EVA and they were using up more oxygen than anticipated and had to call it off 30 minutes early. They had travelled about five miles during the first day on the Moon. Their exploit was front page news. "Rocky Ride for Moon Motorists"[12]; "Rock 'n' roll drive on the Moon"[13].

Falcon *performs the first live lunar lift-off.*

The highlight of the second day on the Moon came when Scott discovered what has been described as the Genesis Rock. "Oh man, guess what we've found. I think we might have what we came for". The crystalline rock was thought, uncorrectly as it turned out, to be primevial lunar material. Scott and Irwin, a Colonel and Lieutenant Colonel in the US Air Force respectively, were identifying rocks and describing them in a highly professional way, testimony to the many months of training that they had in geology, from none other than Jack Schmitt.

The ever-watchful remote camera spotted that Scott had left a sample bag behind and Houston was able to remind him to pick it up. It was a not just case of Big Brother watching but listening, too, as Joe Allen's loud "tut, tut, tut" reminded Scott to watch his language after he said, "Oh shit, I forgot the camera. Shit!" One newspaper describing the incident, reminded us disapprovingly that Scott was "a colonel of the United States Air Force". The astronauts fell over a number of times. One occasion was particularly dramatic. Irwin tipped over on a 20 degree slope and ended up on his back pack, flailing like an upturned insect. The day ended with a lovely deadpan remark. As he parked alongside *Falcon*, Scott said, "Home sweet home. That was a nice trip". The nice trip was described by the next set of headlines: "Apollo 15 Moon find delights scientists"[14]; "The four letter slip up!"[15]; "NASA jubilant over Apollo Triumph"[16]; "Apollo Drops 4-letter word"[17].

"It's fair to say," said NASA scientist John Salisbury, "even now, their mission has been more successful than all three previous landings together". He was describing events after the third day on the Moon during which the astronauts continued to react with unrestrained glee at their fresh discoveries. Getting tired and tetchy towards the end, when Scott was drilling a reluctant core sample, the commander snarled, "Tell me, do you really want it that bad? I don't think it's worth doing. I'll tell you, you're really investing a lot in this thing. The crew will break down, the stem never will. How many hours do you want me to spend on this drill". Turning to Irwin standing alongside, he said, "You could be doing something useful instead of just standing". When he finally got the drill bit apart to release the core sample, Houston asked how he managed it. Scott replied in sardonic fashion, "I took it apart." His tetchiness could be explained by his extremely sore fingers, which he revealed after the flight. There were some pleasant moments for the urbane colonel, however, particularly what he called the "Galileo Experiment". In a televised demonstration aimed at students, he dropped a hammer and a feather simultaneously. Simultaneously they hit the surface. "Hey, Galileo was right," chorlted Scott, scuffing dust over the feather which he had

Farewell view of Hadley Rille.

planned to bring back to Earth. He never found it.

For the first time, TV viewers were able to watch a lunar lift-off live. The camera mounted on the parked lunar rover showed Falcon preparing for lift-off. "Okay troops," said Allen, "explorer hats off and put on the pilot hats." The crew responded with, "we are sure ready to do some flying". And fly they did, into the wild black yonder. At the moment of the multicoloured, sparkling lift-off, Worden, another Air Force officer, in *Endeavour* played the US Air Force music, 'Wild Blue Yonder', for everyone to hear. As *Falcon*'s ascent stage made its way towards *Endeavour*, Houston said, "We monitored you at lift-off and can confirm that you lifted off." Scott replied, "That's good to know". He was in a good mood once more. *Falcon* approached, Worden docked and said "Welcome home". The crew had to remain in pressure suits for a while longer because there was an unusual rise in pressure in the docking tunnel. *Falcon* and a small sub-satellite were deployed before *Endeavour* lit its engine for home.

Worden, the quiet and forgotten man on the mission, had his chance for glory: the first deep space EVA, some 206,000 miles from Earth. He spent about half an hour retrieving two cassettes of film from mapping cameras. What he was not able to photograph himself, and what he later described as a "fantastic sight", was Irwin standing in the open hatch with the full Moon behind him. "Worden's 2,000 mph Moon Stroll"[18].

On the way home, the crew described their feelings during an in-flight press conference, justifying vigorously the cost of the mission and denying that they had been pushed too far on the Moon. A surprise was in store for them at splashdown. As the three giant chutes deployed, one failed. This was spotted and the information was relayed to *Endeavour*, "Be prepared for a hard impact". A safe splashdown and recovery followed, then it was "Apollo Tames the Moon"[19] and a tickertape drive in New York. There were problems, however. First the astronauts' hearts appeared to have been weakened by the moonwalks more than previous ones. Irwin complained that he suffered dizzy spells on the Moon. Secondly, it was later reported that the three heroic Apollo 15 crew had been unceremoniously grounded, for franking envelopes on the Moon and selling them on behalf of a trust fund for their children. When the unmanned Luna 18 crashed on the Moon soon after Apollo 15 got home, it seemed to answer all the questions about the value of having man in space. The crew, it appeared, had simply displayed the normal frailties that go with being human.

Apollo 16

44th manned spaceflight
16 April 1972
11 day 1 hr 51 min 5 sec
71 hr 14 min on Moon
213 lb of Lunar samples

The Jokers of Descartes

NINETEEN SEVENTY-TWO BEGAN WITH President Nixon giving a go-ahead to the Space Shuttle programme. It would cost $5 billion to build and the first launch would be about 1978, Congress permitting, of course. "NASA is under considerable pressure to cut its budget, so it is considering using solid rocket boosters, rather than more expensive liquid-fuelled ones, to propel the space taxi"[1]. How this would effect the safety of the Shuttle would be clear to see 14 years later. Despite the budget threats, Wernher von Braun felt able to predict confidently that a baby would be born on the Moon in the year 2000, such was the euphoria that the successfull Moonlandings had created.

Apollo 16 was rolled out to the pad in the New Year, which also began with the news that NASA had confirmed that Apollo 17, threatened to be cancelled by budget cuts, would go ahead. Charlie Duke, the Apollo 16 lunar module pilot was safe. Then the guy caught pneumonia. The launch was set for 17 March. Fortune smiled on Duke, because difficulties with the command module-lunar module separation system meant that Apollo 16 was delayed to 16 April. Charlie Duke, with time to recover, was saved yet again. An earlier bout of ill health had caused Ken Mattingly to be dropped from Apollo 13. The Duke-Mattingly link was perpetuated by Apollo 16. These two rookies had as their commander, the taciturn, dead-pan John Young. The trio were an unknowned quantity in terms of their ability to brighten up the flight for dwindling TV viewers and spacewatchers. It could be a quiet mission, some thought. How wrong they were. The build-up to Apollo 16 was pretty non-existent, reports about the vehicle being returned to the vehicle assembly building for repairs, apart. Reportage centred on other events: "By Jupiter, there's a Pioneer on the way!"[2] as Pioneer 10 began its journey into infinity; "Moonrock for Moscow"[3], as Luna 20 came back with a few grammes of soil; "Russian plan for manned Mars satellite in 1978"[4]; "NASA cancels Nerva nuclear-powered engine"[5] – due to budget cuts". Nearer to Apollo 16's lift-off we were told that "Moon men can get divorced"[6]. Young had divorced and remarried while training for Apollo 16, the first astronaut to do so. Ken Mattingly's wife was expecting a baby and he was prepared for the nervous, long float up and down the spacecraft.

Before launch, Deke Slayton spilled the beans on the apparently luxurious pre-launch 'breakfast' of filet mignon and all the trimmings. It was was actually a "TV-type steak and eggs dinner and not very good," said Slayton, "one of the penalties of flying a mission". The launch was "right down the middle of the plot board," said mission control and once in orbit the usually dour taciturn Young could not contain himself.

"The thing worked like a gem.... Boy, it's just beautiful out here looking out the window, it's utterly fantastic ... right on ... we're coming into darkness and the sunset is beautiful as it always is in the space business...."

The following day, after a safe TLI and TDM, command module *Casper* and lunar module *Orion* were chuntering away towards an empty window in space where the Moon would be two days later and the first Apollo hitch occurred: "Debris seen pouring from Apollo craft"[7]. "Apollo speeds on with a tail of Shredded Wheat"[8]. Young reported that a cloud of brownish particles was streaming from a vent in the lunar module. They seemed to be coming out with some force, he said. Duke described the event as pieces of material 'shredding' off the craft, looking like 'shredded wheat'. Mattingly chipped in that it reminded him of "paint off an old barn". That's pretty much what it was. The alarm was short lived. A panel on the side of the lunar module had been treated with white silicon paint, as a precaution, in case the moon landing was delayed a day and the sun angles would be more critical to the sensitive surface.

The next problem sounded an alarm in the cockpit, startling the astronauts from their sleep period. *Casper*'s guidance system had malfunctioned as a result of electrical interference. The crew were sent back to bed while controllers checked the matter and were woken one hour early to be told that all was OK. Young wasn't cross at being stirred from slumber yet again, "That's the best news I've heard today," he said. As the flight motored on, mission control heard "Allo. Allo. Allo...." coming from Apollo. Then a Spaniard singing. Mystified they accused Young and Duke of a bit of leg pulling. But Young and Duke were equally mystified. It turned out that the fourth 'crew member' was a Spanish telephone engineer up a pole near the Madrid tracking station who had inadvertently been patched over the ground circuit. The engineer was 'electronically isolated'.

"Hello, Houston, Sweet Sixteen has arrived," said Young as Apollo 16 emerged from the lunar farside after the LOI burn. "It was a super double-fantastic burn, that baby just rifled us down the line". Was this really the taciturn Young speaking? Later, *Orion* separated and *Casper* prepared to back off. It all seemed so matter of fact. Suddenly: "Fear of Apollo Disaster"[9]. In fact, the astronauts had experienced several checkout problems with *Orion* after undocking but, at last, all had seemed clear. Then *Casper*'s engine failed to ignite in order to take Mattingly into the parking orbit, to await *Orion*'s return. This turned out to be a little matter of a fault in the back-up yaw gimbal drive servo-loop of the service propulsion engine. If the engine could not be ignited, Orion would have to redock and its engine would have to be used to take the astronauts out of lunar orbit instead of down to the Moon. This was a potential Apollo 13 situation but it seemed so unreal because there did not seem to be any urgency about it. This was because the fault lay with a back-up system. Mission rules dictated that both primary and back-up systems had to be working before the critical phase of the mission proceeded. Thus, if the back-up system could not be made functional, the landing was off. The seconds were ticking by. Soon it would be impossible to land at Desacartes anyway because the it would be out of range. Duke said, "if we ever get to land this thing, we would like to sleep first before the walk". As he passed over Descartes, Young described the surface he was hoping to attempt to land in later, "a big craggy cinder field with house-sized rounded clinkers". "It's Still go-slow on the Moon"[10] said an early edition. After a nerve tingling three and a half hours, the fault was rectified. "You do have a go for another try," said mission control. "I'm all ears," said a joyful Duke. "Go, go after Apollo hitch"[11]; "1am: It's go for Moon landing after hitch"[12]; "4am: They're down safely on the Moon. Apollo Touch – and Go!"[13]; "Moonmen men safe after six hour agony"[14]. It didn't sound like agony to us, more like ecstacy.

The high-spirited astronauts had sounded like they were on a roller coaster ride as they came down to Descartes. "Coming in like gangbusters. I can see the landing site. We're right there, John. Man, there it is, Gator, Lone Star. Perfect place over there, John. A couple of big bolders, contact, stop. Wheeee.... Orion is finally here, Houston. Fantastic!" Young let out a loud "whoopee!" The ecstatic crew continued a light-hearted, high-spirited banter. "All we have to do is to jump out the hatch and we've got plenty of rocks! I'm like a little kid on Christmas Eve," said Duke. " I really want to get out," said Young, looking forward to his Moonwalk, "but I think discretion is the better part of valour".

The crew had been the first to have a drink during the descent, from a small dispenser with a straw inside their helmets. The orange juice was laced with potassium because doctors felt that this would help the astronauts' hearts to function better during the arduous Moonwalks. Duke's pack burst and he was showered with orange juice. "I wouldn't give you two cents for orange juice as a hair tonic." This was not the last time that orange juice would feature in the mission. Back on Earth, a jubilant Dottie Duke said that she and her husband, "both get kind of excited". As for the Moonwalk, you were either going to like it or cringe with embarrassment.

Young came down the ladder, stepped onto the surface of Descartes and slowly looked around the scene. He was getting the explorer bit right. Then he said, "There you are, mysterious and unknown Descartes and Cayley Plains." So far so good. "Apollo 16 is going to change your image". Well, it was worth a try. Duke brought things back to normal when he jumped onto the surface. "Hot dog, isn't this great!" Within an hour, the astronauts had unfolded their rover and were setting off for some exploration. They observed that *Orion* had landed in just about the only smooth spot around. Anywhere else and it would have landed on a slope, said Duke. Young felt that if he let the rover free, it would roll back under the lunar module. The TV camera was switched on and at last viewers could see the astronauts with amazing clarity. The first steps could not be transmitted because of antenna problem on *Orion*. As they worked away, Duke was so happy, he chortled, "aw, aw, aw, this is super!" The flag ceremony was staged to perfection.

While posing for a picture by the flag, Young was told by mission control that Congress had cleared the way for the development of the Space Shuttle. The patriotic Young said, "the country needs the Shuttle mighty bad". Nine years later, almost to the day, he would be commanding its first flight.

Standing by the flag, Young was urged by Duke to jump into the air and give him a "big Navy salute". This Young did, giving a vivid demonstration of what can be done in one-sixth g. Duke threw a steel bar. "I threw it 200 feet or more – I'm going into the Olympics hammer event," he laughed. The moonwalk also featured a Grand Prix on the moon as Young, urged on and photographed by Duke, really put the rover through its paces, churning up piles of moondust and at one time flying through the air with all four wheels above the ground. Young reached a record speed of six miles per hour during his spin around a 500 foot triangular circuit.

One of the most important instruments to be placed onto the surface concerned the heat flow experiment, worth about a million dollars and a lot more to scientists. Duke patiently drilled a ten foot deep hole in the Moon to drop a sensor into it. The sensor was attached by a cable to the experiment housing. Millions of viewers saw Young walk

Young takes the rover for a Grand Prix spin.

Duke's great hole in the ground.

Young the explorer.

clumsily towards the cable and rip it out of the experiment. The crestfallen Young said, "Charlie, something happened here." Duke replied, "What happened?" "Don't know, it appears a line pulled loose.... Oh, God Almighty, it broke right at the connector." Duke, sounding slightly cross, said, "I'm wasting my time". Damn, I'm sorry," said Young. Scientists were bitterly disappointed. "It is a sad blow," said Dr Donald Beattie, "it was the highest priority experiment". It was this episode, combined with the high jinx of the two test pilots of Descartes, that gave an incongruous feel to the Apollo 16 mission. Would a scientist have acted like a little kid on Christmas Eve? some asked. The newspapers treated it all lightheartedly, too. "Moon men play it for laughs"[15]; "Moonmen take a joy ride"[16]; "Ooops! Young trips us and wrecks Moon experiment"[17].

After seven hours on the Moon, the crew retreated into *Orion*, thinking that they would have difficulty getting to sleep in the relatively uncomfortable abode. A good night's sleep afterwards, however, and Young was telling mission control, "I couldn't believe it. This guy Charlie sleeps like a babe". In a letter to Keith Wilson, a manned spaceflight historian and researcher, Duke described the sleeping arrangements in the lunar module: "We had two small beta cloth hammocks which we attached to the spacecraft. After taking off our spacesuits, we just lay down in the hammocks and got a good night's sleep. John's hammock was in the upper portion of the lunar module. The foot of the hammock attached to the sides of the instrument panel just below the telescope or sextant. The head of the hammock went to the rear of the lunar module about 18 inches above the top of the ascent engine cover and attached to the rear bulkhead of the ascent stage. Below that on top of the ascent engine cover were out two spacesuits. My hammock was perpendicular to John's and was in the foot well of the lunar module. My feet were basically in the position where John would stand at the commander's station and the foot of the hammock attached to the left bulkhead of the lunar module about six inches off the floor and the head of the hammock attached to the right bulkhead and my head as I lay in the hammock was basically where I stood at the lunar module pilot's station. This was very comfortable, though it was tight quarters in the lunar module. I found sleeping in one-sixth gravity not much different from sleeping here on earth except you don't have to move as much to maintain proper circulation."

The next Moonwalk featured a two and a half mile ride to the foothills of Stone Mountain, 700 ft above *Orion*. The crew also visited a young crater called, not surprisingly South Ray Crater, where they collected samples that were strewn in rays over the surface from 500 ft within the crater after a meteorite impact millions of year previously. When they had driven to North Ray crater, 1,200 feet deep, Duke exclaimed, "Man, is that a hole in the ground!" The rover, churning through dust that was as heavy-going as a ploughed field after rainfall, gave Young "the roughest ride I've ever been on", such a rough ride that it broke the vehicle's yaw degree indicator. Towards the end of the second walk, the crew was urged to get back inside *Orion*. "Why don't you give us an extension?" asked Young. Duke chipped in: "How about an extension, you guys? We're feeling good." Young said, "All we were going to do tonight was sit around and talk". Mission scientist, Tony England replied," we like to hear you talk". Duke pleaded: "ten minutes and we can get all this done. How about ten minutes, Tony please. Come on Tony, please". England relented, "Okay. we'll give you 10 minutes, How's that? Just because we love you".

The rover wasn't the only thing giving the crew a rough ride. So was the potassium-laced orange juice. Back inside Orion, Young described the symptoms vividly to Duke. "I got the farts again. I got 'em again Charlie. I don't know what the hell give them to me. I think its acid in the stomach, I really do. I mean, I haven't eaten this much citrus fruit in twenty years. And I'll tell you one thing, in an another twelve f...... days, I ain't never eating any more. And if they offer to serve me pottasium with my breakfast, I'm going to throw up." Young, who was brought up in the Florida orange belt, around Orlando, continued: "I like the occasional orange, I really do. But I'll be damned if I'm going to be buried in oranges". Mission control quietly interupted this live conversation, with "Did you guys know you got a hot mike?" "No-o-o!" replied Young, "How long we had that?" "It's been through the debriefing," replied mission control, no doubt holding back the sniggers. "Oh," came the reply from *Orion*. Meanwhile, mission control was deciding to shorten the third moonwalk from seven to four hours and to bring the crew home a day early, because of continued worries about the *Casper* service propulsion engine. Programme manager, James McDivitt, explained, "The systems in that engine are not as good as they were at lift-off. The longer you stay up the greater the risks". The third moonwalk began with the sun high in the sky. "Charlie, it's gonna be hot out here today," said Young. While the temperature was to soar to 180°F, the water-cooled underwear and portable life support systems of the astronauts kept them at a comforable 70°F or so. Driving down a slope towards *Orion* for the last time, the rover reached a record speed of 10 mph. Claiming a world record, four and a half mph higher than the design limit, the crew was reminded to be careful. The astronauts certainly seemed to be taking risks during their final walk, when they indulged in a little high jump competition which Duke won outright but then landed heavily on his portable life support system. It seemed an extraordinary thing to do. "The lunar olympics," Young announced. "We're gonna show that a guy could do, like jump flat-footed straight in the air, three or four feet". When Duke fell, Young said rather

Orion *returns to* Casper.

self rightously, "Charlie, that ain't very smart". "Sorry about that," said the chastened Duke. They then entered *Orion* for the last time, leaving the rover parked nearby, to show the world another lunar lift-off. Meanwhile, Mattingly had quelled some worries, firing *Casper*'s engine to place it in an orbit to receive *Orion*'s ascent stage.

"What a ride, what a ride!" exclaimed Young, as *Orion* took off amid a multi-coloured shower of sparks. After docking and the ejection of a sub-satellite, the service propulsion engine burned successfully, taking the crew out of lunar orbit and raising their morale "a couple of one hundred per cent". Mattingly, who had flown the longest solo spaceflight in US manned space history, performed a deep space EVA later, which was a complete anticlimax except perhaps to him and his expectant wife. Lifting his visor for a moment to inspect closely his work outside, Mattingly was reminded by mission control, watching on TV, of the danger to his unshielded eyes of the Sun's ultra-violet rays.

A remarkable incident during the EVA was later recalled by Duke. Mattingly had taken his wedding ring off during a previous experiment inside the command module and had lost it. During his EVA, with Duke standing with his helmeted head out of the hatch, Mattingly's ring floated out of the cabin just eluding Duke's grasp. It floated towards Mattingly, bounced off his helmet and flew straight back to Duke, who caught it!

After his exhuberant work on the Moon, Young rather amusingly put down the scientists who were eager to learn about his lunar spoils. "It's too soon to be making any conclusions about the region," he said, "it ain't good science". As the flight was coming to an end, Young said, "I think we have seen more than most people could in ten lifetimes". Back at a welcome from space centre workers at Houston, Young looked up at the Moon and told them, "well, here we are, and there it is. It's hard to believe we were up there".

Casper *from* Orion.

Apollo 17

45th manned spaceflight
7 December 1972
12 day 13 hr 51 min 59 sec
74 hr 59 min on Moon
243 lb of Lunar samples

The End at Taurus Littrow

BY THE TIME CHARLIE DUKE HAD BEEN assigned to Apollo 16, Apollo 17's lunar module pilot, USAF test pilot Joe Engle, knew he had lost the Moon to Jack Schmitt. Selected in 1966, Engle was thought of as one of the most likely to succeed. A former pilot of the X-15 rocket plane, he had soared over 50 miles high, three times, qualifying him for USAF astronaut wings. In 1964, he was named as one af America's outstanding young men. Robbed of the Moon, Engle missed out on the Skylab space station and wanted no part of the Apollo-Soyuz adventure. Five years on, he was commanding the *Enterprise* Shuttle glide tests in the atmosphere. In 1981, he commanded the second Space Shuttle test flight and three years later took another Shuttle to the skies.

Resigning from Nasa after the *Challenger* disaster, Engle became an aerospace consultant. So he didn't do too bad. But he will always be known as the man who lost the Moon. Many people sharing Engle's loss forgot that his inevitable replacement Jack Schmitt, was a trained pilot who had been selected a year before Engle. There were of course, equally as many people who thought that it was about time that a real geologist set foot on the Moon rather than test pilots pretending to be.

By the time December 1972 arrived, the Engle episode had been forgotton and attenion was being focused on the last manned moon trip and why it was the last. "What We Got For $10,000 million"[1]. Basically nothing that could not have been brought back by unmanned craft, more cheaply, said most poeple. "The great American assault on the Moon is almost over. There is to be no follow through. Man is withdrawing from Moon exploration, not because of the hostility of the lunar surface but because of indifference back home," wrote the *Sunday Times*'s Bryan Silcock. "There is little chance of anything comparable to the Apollo programme ever being undertaken again. The circumstances that produced this technological orgasm were unique ... but however disappointed politicians and soldiers may be with the results, there is one group of people that has harvested a rish crop from the barren surface of the Moon: the scientists. It is an ironical outcome, for in the Apollo programme, science has always taken second place to engineering"[2].

Why was it the last? Angus Macpherson of *The Daily Mail* wrote, "The answer must lie in the psychology – not only of the 200 million Americans that paid $50 a head to let their boys go to the Moon – but of the rest of the watching human race whose gasp of amazement had frozen into an irritable yawn in four short years, since the first Christmas journey when Apollo 8 visited the Moon"[3]. Even Wernher von Braun himself conceded that "the awareness of the Dream had gone sour, and a great journey had petered into something of a dead end". Von Braun saw it as a "simple dollars and cents affair, we have achieved that goal we set for ourselves. It's time now to prove that we can provide an economic return from space and use it to solve some of Earth's problems". To sum it all up, Boeing workers went on strike at Cape Kennedy, threatening to scrub the Apollo 17 launch. The lift-off was saved when workers were reportedly given a 40 per cent pay increase! Those were the days.

"In all the 4,600 million years since its birth, the Moon has not been subjected to such thorough physical examination as it will get from ... Harrison Schmitt," said *The Daily Telegraph*. "I cannot say specifically what I can do that someone else cannot do," said Schmitt before taking off, "but I can apply my experience to observe geologic and other scientific features on the Moon. I think it will pay off to put someone with that experience on the Moon". The grey haired, 38-year-old-commander of the mission, Gene Cernan was under a bit of pressure here. Who was boss on the moonwalks? Cernan, perhaps a little niggled, said that he thought he was just as good a geologist as Schmitt, just as he felt Schmitt should think himself a better pilot. More publicly, he was to say, "Jack's professionalism in the field of science and geology can only enhance our capability once we get there". To which Schmitt commented, "I'm human enough to think I can do a better job, at least in this case, in the area of scientific exploration".

Also flying with them would be command module pilot, a balding jack-in-the-box, Ron Evans, an old buddy of Cernan during their Navy years together. Both got plastered the night Cernan got accepted by Nasa and Evans didn't. Apollo 17 had other crewmembers: five pocket mice from California. The astronauts knew the risks but felt confident that they wouldn't lose their lives, which is more than could be said of the mice, which were to be killed after the flight to determine how cosmic radiation had affected their brain tissue.

Cernan blew a farewell kiss to his daughter Tracy and set off for the launch pad. While he encased himself inside Apollo, Tracy set off to watch the launch together with thousands of other people, including Dr Rudolph Nebel, (the man who built the Nebelwerfer surface-to-surface rocket used by the Wehrmacht during the Second World

Taurus Littrow awaits.

War), and 130-year-old Charlie Smith, a former negro slave brought to the USA in chains 118 years before. His 70-year-old son was also there. Showbusiness representative among the 500,000 veiwers were Bob Hope and Frank Sinatra.

The launch was going to be one with a difference for it was the first to take place at night – in the US manned space programme – but the intensity of the mighty Saturn's flames were expected to turn night into day for miles around. The count went well. At T-30 secs and counting, observers clearly saw a red flash appear at the base of Apollo. This was probably due to a build up of gas just prior to the ignition sequence starting and was unconnected with the fact that the count stopped. "We have cut off," said the man reading the count. There was intense disappointment, as it was discovered that the terminal countdown sequencer had spotted that the liquid oxygen tank on the third stage had not pressurised. This was attempted manually but the sequencer refused to accept. At first it was thought that the crew would have to evacuate the command module and use the emergency slide wires to get into underground bunkers for safety but this thought subsided when the reason for the stoppage had been discovered. As Cernan suggested that the crew "start a nice conversation about a good book", technicians worked feverishly to get a lift-off. The final

Schmitt, the flag and the Earth.

Schmitt and the rover.

Apollo flight was the first to have problems so close to lift-off.

The fault was rectified and the count-down began again at T-22 minutes, just 36 minutes before the launch window ran out. Viewers would still get their night-time lift-off and Apollo would get to the Moon on time. "Seventeen is go!" exclaimed Cernan as he rose above the firey scene below the arcing Saturn 5. The launch was awe inspiring, indeed turning night into a mid-day and a sunny one at that. "You're right down the pike, 17," said cap com Bob Overmyer. Cernan felt the ride was a good one but "it is rumbling around a little bit". Apollo 17 reached orbit and later was despatched on TLI. Transposition and docking followed and 14 hours after boarding Apollo, the crew got a well-earned six hours' worth of sleep. After waking, a cheerful Cernan translated wiping the sleep out of his eyes as "give us a few minutes here and we'll get operational". The first newspaper reports were being published. "Drama – then moon men go"[4]; "Smooth trip after Apollo hitch"[5]. One report chose to concentrate on the fate of the Kennedy Space Centre after Apollo: "26,000 jobs down to 15,400 and counting"[6].

The astronauts flew on, with Schmitt being increasingly intruiged watching the Earth's ever-changing weather patterns. "You're a regular weather satellite," said a cap com. The translunar coast, or TLC, was as smooth as silk and the crew even overslept one 'night' after taking sleeping pills. They slept through an assortment of sounds blasted to awaken them, including a noisy football chorus and a klaxon horn. "Apollo crew sets Sleep-in Record"[7]. It took an hour to wake them up. "we're asleep," said a drowsy Cernan. "That," said cap com Gordo Fullerton, "is the understatement of the year". Cernan joked, "that was some party last night, it was a real humdinger. Our biggest problem this morning is keeping Ron from going back to sleep".

Before the lunar orbit insertion burn, Cernan had a session with the doctor because of a problem of too much gas in his stomach. The conversation was a private one. Going public again, the second-time entry into lunar orbit for Gene Cernan was "the smoothest and quietest I can remember". Soon after, Schmitt, the geologist, thought he saw a meteorite impact near the crater Grimaldi, spying quick flash of light in the Earthshine. Shortly after, Apollo's S4B stage, which had been following the men to the Moon, hit the lunar surface near the crater Ptolemy, causing a moonquake for the benefit for the seismometers already laid out by previous astronauts. *Challenger*, with Cernan and Schmitt aboard, separated from *America* and began its controlled plunge towards the Moon, a small target at Taurus Littrow, hemmed in on three sides by 6,000 feet high mountains.

The tone of the conversation was as usual, short on science, high on euphoria. "I can even see where we're going to park this baby down.... Come on baby.... Are we coming down, oh baby ... 200 feet ... feels good ... very little dust, very little dust ... stand by for touchdown ... 10ft ... engine stop." Challenger fell the last ten feet. "Contact, that was a real meatball". Cernan eventually formally announced the arrival, "OK, Houston, the Challenger has landed." A frenetic conversation then ensued, "We is here. Man we is here.... Beautiful all the way.... Nice boulders out there ... Hey, look at that rock.... Absolutely incredible.... I think I see the rim of Camelot (a crater).... Oh, my golly, it's unbelievable". Schmitt commented, "this is the most majestic moment of my life. This is something everyone's gotta do in his life". He then rankled Cernan, who had landed with plenty of fuel left. "We could have gone all round, we could have hovered a while," Schmitt said light heartedly. Cernan seemed to cut him dead. "I like it right where we are," he said. Schmitt, rubbing the salt in a bit, said, "Who told you this is a flat landing site?" Cernan replied, "What do you want, an absolute guarantee?"

The coverage of the last manned landing on the Moon, said it all. "Whoops – they're on the Moon"[8], said a 36 line story on the front page of one newspaper; "Apollo lands at walls of Camelot"[9], said another. "Bulleseye on the Moon"[10], reported Angus Macpherson of the *Daily Mail*, reduced to just 36 lines by the editor's pen.

The first walk on the Moon was heralded by one newspaper headline which emphasized the difference between the two last Moonwalkers perfectly, "Scientist and astronaut step out on Moon after perfect landing"[11]. Cernan, a mere astronaut and seemingly taking second place, and wearing a helmet with a fabric nosescratcher in side it, stepped first on the Moon. "OK, Houston, as I step out on the plains of Taurus Littrow" ... Yes? ... "Apollo 17 is ready to go to work." ... Oh. The seven hour walk was an exhausting effort. The drill got stuck. Cernan drove the wrong way in the rover. The car broke down. The left rear fender came off, showering the astronauts with soil. Cernan had to stop to make repairs. "Makes me sort of mad," he said. Schmitt was earning praise for his professional descriptions of what he called 'a geologist's paradise' but spoiled things by lowering the tone with astronauts' banter. Asked to take another look homewards, he said, "When you've seen one Earth, you've seen 'em all". Then there was a real cringe job, a song. "Walking on the Moon one day, in the merry, merry month of May.... " Cernan cut in, "December!". And so it continued. The newspapers lapped it up with what coverage there was. "Not all play in the lunar laugh in"[12]; "Lunar dust up hold up"[13].

The second moonwalk could have been one of the most historic moments in history. Schmitt happened across some strange coloured soil. Kicking it up he exclaimed, "Oh, hey, there is orange soil!" The astronaut told the scientist, "Don't move until I see it!" Cernan arrived to inspect the soil, "It's orange, Wait a minute, let me lift my visor up, it's still orange!" The excited Schmitt replied, "Sure it is, I gotta dig a trench Houston. Zap me with a little cold water. Fantastic, sports fans. It's trench time". Sounding more like a geologist again, he said, "Having all the colour changes and everything, I think we might consider we have a volcanic vent. I'm not sure how we prove it." Listening to comments from Earthbound geologists, excited beyond their wildest dreams, one could be forgiven for thinking that Schmitt didn't have anything to prove. According to mission scientist Dr Farouk El Baz, "It told me we've found volcanic activity and perhaps even water vapour within the moon". Dr Stuart Agrell of Britain said, "This is the most exciting discovery of the Apollo 17 mission. There can be now be no doubt that water existed on the Moon". The newspapers thought so too. "Dig that crazy Moon soil: It's orange!"[14]; "Now the search begins for Moon water and lunar life"[15]. Apollo 18's grounded commander, Dick Gordon must have pricked his ears up when he heard some suggestions that his mission may be saved. "Orange dust may save Apollo 18", reported one newspaper[16]. The hope was short lived. Wernher van Braun said, "Apollo cannot be revived. Most people have left and we have passed the point of no return. But once we have our new spacecraft, the shuttle, it will be possible to go to the Moon more economically". Things seemed to be getting a bit out of hand, especially when, in response to Schmitt's exclamation "momma mia" during the second walk, Italian newspapers went wild. "Italian spoken on Moon!"

The third walk ended with elongated ceremonies and speech making to match the mood of the occasion. The astronauts unveiled the plaque mounted on the lunar module *Challenger*: "Here man completed his first explorations of Moon, December 1972. May the spirit of peace in which we came be reflected in the lives of all mankind". Cernan said, "this is a commemoration which will be here until somone comes back to read it again to further the meaning of Apollo". Schmitt proclaimed, "I'd like to remind everybody that this valley of history has seen man complete his first evolutionary steps into the Universe, leaving the planet Earth and going forward into the Universe. I can think of no more significant contribution that Apollo has made to history. Erecting a flag that had flown in mission control during all the previous flights, Cernan addressed the younger generation: "The door is now cracked but the promise of the future lies in the young people, not just in America, but in young poeple all over the world learning to love and learning to work together." Holding a rock towards the TV camera, he continued, "this rock is typical of what we have here in this valley. It's a rock composed of many sizes, many shapes, many colours, probably from all parts of the Moon, that have come and grown together. We'd like to share pieces of this rock with many countries of the world, in the hope that this will be a symbol of what our feelings are, what the feelings of the Apollo programme are and as a symbol of mankind, that we can live in peace and harmony in the future". Cernan thanked the thousands of American workers who had made Apollo 17 possible. He aslo thanked God. "If he's listening, I'd like to thank him too". Then came the final, moving departure speech. With the TV camera mounted on the rover close by and pointing at *Challenger*, with Cernan at the foot of the ladder, the last man to leave the Moon this century said, "As I take these last steps from the surface, back home for some time to come, but we believe, not too long into the future, I believe history will record that America's challenge of today has forged man's destiny of tomorrow. And as we leave the moon and Taurus Littrow, we leave as we came, and God willing, as we shall return, with peace and hope for all mankind. God speed the crew of Apollo 17". The curtain call should then have followed.

After a night's sleep, however, the crew gave us an encore of a different kind. A duet, *Good Morning to You*, sung to mission control, followed by a rendition by Schmitt of a parody of 'Twas the night before Christmas'. Afterwards, Schmitt quipped, "People always said we ought to have a poet in space". The more serious business of getting backstage, or home, had to follow. To prepare the crew, mission control played the theme from the film, *2001: A Space Odyssey*. "We're on our way, Houston," cried Cernan as he took off in a shower of multicoloured sparks. "Challenger is go!"

Docking with *America* and the quiet Ron Evans was a dream. The top half of *Challenger* was sent plunging towards a target about ten miles from its bottom half at Taurus Littrow. Evans thought he spotted the impact crater. The seismometer certainly recorded the impact but the TV camera on the rover did not, as it had been hoped. The dirty Moonwalkers messed up Evans's home and the vacuum cleaner came out. This still didn't get rid of all the slightly tacky moon dust. Before the big burn to send them out of lunar orbit, the crew awoke to hear from mission control, the song, *Come on baby light my fire*. The burn was a good one and later, *en route* to home, Evans performed the now customary spacewalk.

The re-entry targeting was excellent, so good, in fact, that mission control sent a telegram to the tracking ship telling it to move 50 feet out of the way! The astronauts were hoping that a lost pair of scissors would miss them if it fell out of its secret hiding place during re-entry. It could hit with the force of an axe in seven gs. "Space men hunt deadly dagger"[17]. The scissors stayed put and Evans reported, as *America* descended beneath its three stripped chutes, "It's a beautiful day, all's well on board". In no time at all, the crew were bouncing along the deck of the recovery ship, the *USS Ticonderoga*. Cernan was first to the microphone, "By golly we're proud". The crew returned to a tumultuous welcome at Houston, and within a few days it was business as usual.

"Luna 21 ends Russian's space rest"[18]; "NASA plans for Mars under cloud"[19]; "Doctor sues over space monkey trip"[20]; "New space lab goes into orbit"[21]; "Second probe to Jupiter"[22]. However, for the next thirty years at least, there would not be these headlines again: "Man on the Moon".

The final farewell to the Moon.

The end of The End.

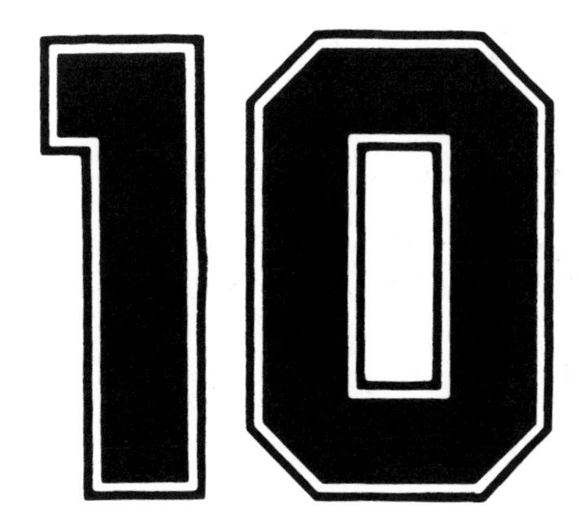

The Twelve:

THE APOLLO MANNED LANDINGS ON THE moon were perhaps the greatest human and technological feats achieved in history. The Apollo programme was an immense fusion of human resources and technology, the ultimate *team* effort. The twelve men who walked on the Moon were always the first to remind people of this fact. They were figureheads in the vast enterprise: the point at the end of the rocket. The human interest in them as the astronaut heroes was to them a 'non-issue', a favourite expression among their exclusive cadre at Houston. Most of the ardent public learned about them through the media and to most of the astronauts the media coverage about them and about their flights was 'crap', concentrating on the non-issues. Who cares who got out of Apollo 11 first? Who cares if one of the Apollo 12 moonwalkers actually hummed and whistled on the Moon? Who cares that the Apollo 14 commander fought his way to the Moon by beating an illness that grounded him? Who cares about the Apollo 15 envelopes? Who cares about the larking about by the Apollo 16 crew at Descartes? And who cares that a geologist bumped a test pilot off the Apollo 17 crew? These questions the astronauts ask. "It could have been any of us, it just happened that we were in the right place at the right time", they say. To the astronauts, these are non-issues. They would have us concentrate on the real issues: the technological achievements by faceless professional, engineering 'Right Stuff' test pilots. But Apollo achieved the first – and so far only – manned landings on the Moon. That man, not a machine landed on the Moon, makes the men who made the landings human beings which fellow human beings are interested in knowing about – as human beings. To the human being, not the technocrat, that is the real issue. The twelve didn't set out to go to the Moon in order to come back to travel the world sharing their experiences with all and sundry. They did a job and had careers to continue. All the same, if only all of them would admit it, those twelve who walked on the Moon have an important tale to tell the rest of us. So we perhaps should be grateful that at least some of the twleve have set aside part of their lives to share their unique human experiences with their fellow human beings on the Earth.

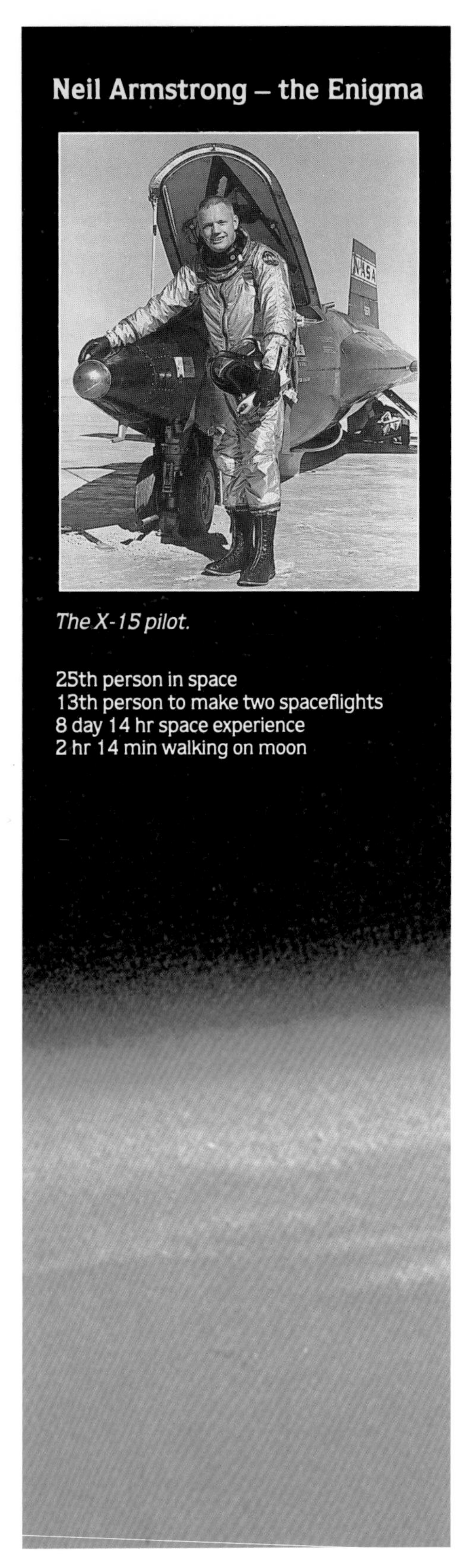

Neil Armstrong – the Enigma

The X-15 pilot.

25th person in space
13th person to make two spaceflights
8 day 14 hr space experience
2 hr 14 min walking on moon

NEIL ARMSTRONG WAS A TEST PILOT. HE flew an engineering test mission. The mission just happened to make the first landing on the Moon. The objective of the test was to 'land a man on the Moon and return him safely to the Earth'. That men walked upon its surface was incidental. Armstrong did his job. The world doesn't own him because he happened to be the first. It could have been any of the astronauts, who were, in any event, the figureheads of an enormous American enterprise involving thousands of people – and millions of taxpayers. For Armstrong to bask in the glory of being the first man on the Moon would have been obscene. Maybe he could have been a bit more communicative about it but that's not the way he is. He is the person he was before he flew: a shy, reticent man, whose silence means 'no' and whose no is termed an argument[1]. Armstrong didn't portray the dashing image of an astronaut perpetuated by the media in the sixties. He had a pot belly that exposed his penchant for a few cannies and rather rounded shoulders.

Armstrong gave himself to NASA after the flight for the necessary goodwill tours which he graciously performed. That done, he went back to his Ohio roots and slammed the door well and truly shut and woebetide anyone who knocks without an invite, especially if they want to ask about the Moon. During his period as a professor of engineering at a university, if you visited Armstrong in his office, you would never have been given any idea that he flew to the Moon. Typical of the test pilot still inside him was the model of the X-15 rocket plane on his desk. No pictures, no models, no mementoes of Apollo.

Armstrong has been known to give interviews, he even did some TV ads, but generally his reticence about Apollo has been taken to the extreme, certainly to a loathing of the press. A good example of this occured following a sales presentation on Learjet that he gave. According to the journalist Lawrence Wright, when the autograph hunters queued up, Armstrong patiently signed but deflected all attempts to talk about the Moon. Even one elderly veteren of the Gemini programme, who approached Armstrong, almost with tears of adoration in his eyes, was given the polite brush off. "I'll never forget that moment you stepped on the Moon, Neil," said another hero worshipper. "Me neither," said Armstrong!

Neil Alden Armstrong was born on 5 August 1930 on a farm six miles from the Ohio town of Wapakoneta, to Mrs and Mrs Stephen Armstrong. Armstrong Sr was a state auditor, which meant that the family – Neil has a brother and sister – moved frequently, although Armstrong Jr studied for three years at Wapakoneta High School. Young Armstrong built model aircraft and made his first flight in a real one at the age of eight. Well and truly hooked, he wanted to be a pilot when he grew up. At the age of sixteen, Armstrong took flying lessons and paid for them working in a local pharmacy and hardware store. He also worked in a local bakery and, because he was small enough, was the one who had to get inside the mixing vats to clean them. He also turned out doughtnuts by the dozen each night.

Neil was a muscial boy, playing horn in a school jazz combo called the Mississippi Moonshiners. Wanting to go to college to study aeronautical engineering, Armstrong gained a US Navy scholarship to help pay for his course. The story goes that, when told, his mother dropped a jam jar on her toe, breaking a bone. In 1947, he entered Purdue University but within 18 months was called for active duty and trained as a fighter pilot at Pensacola in Florida, with a view to returning to Purdue to finish his course. Then the Korean War broke out and Armstrong headed for the war zone as a 21-year-old rookie combat pilot, to fly Panthers from aircraft carriers. He served in Korea for a year, mainly disabling trains, tanks and buildings. On one strafing run, resembling those depicted in the film *The Bridges at Toko Ri*, the wing of Armstrong's F9F2 jet clipped a wire stretched across a valley. He ejected safely over a US Marine base. On another flight he nursed a badly damaged Panther back to the carrier, The *USS Essex*. With 78 combat missions and three air medals under his belt, Armstrong returned to Purdue in 1952 and finished his course, graduating in 1955. Armstrong joined the National Advisory Committee for Aeronautics, begining a long association with what was to become NASA. He was initially posted to the Lewis Research Centre in Cleveland but, like so many of the astronauts, made his way to Edwards Air Force Base and the NACA station there. In 1956, he married Janet Shearon of Evanston, Illinois after admiring her for three years from a distance. Typically, Armstrong had reviewed the situation very carefully before deciding. At Edwards Armstrong was the epitome of The Right Stuff, throwing himself around the sky in gay abandon. He flew the legendary X-1B rocket plane four times. This was carried aloft under the belly of a mother plane and dropped. Its engine would then ignite. Armstrong's first mission was on 15 August 1957, two months after his son, Eric, was born. His last was the 50th and final flight of the second generation X-1 plane.

Armstrong also piloted the X-5, F-102A and the F5D-1 Skylancer, making simulated landings of the proposed Dyna Soar space shuttle vehicle in 1962. He was also one the legendary veterans of the X-15 rocket plane, half plane, half spaceship, which carried a fellow pilot, the late Joe Walker 67 miles into space in 1963. The first of Armstrong's seven flights was on 20 November 1960. On 20 April 1962, he reached a height of 207,500 feet and on 26 July of the same year, a speed of Mach 5.74.

By this time, the Mercury programme was coming to an end. It was a programme which many of his fellows at Edwards, had regarded with some scepticism, favouring the Dyna Soar concept. Spam in the Can, with the Can being test-piloted by a chimpanzee. Nevertheless, the Moon, the ultimate goal, caught Armstrong's attention, purely of course in a test pilot engineering sense and he applied to join the astronaut corps, becoming the first NASA pilot to become an astronaut, in September 1962, with his wife, Janet one month pregnant with their second son Mark. A daughter, Karen, had died in infancy during their days at Edwards. The family was lucky to survive a house fire soon after they had moved to Houston. Armstrong was lucky to survive his first space mission, too.

He became the eighth of the nine class-two astronauts to be assigned a mission, Gemini 8 in 1966, after serving as back-up command pilot of Gemini 5 the year before. Because of the failure of Gemini 6 to do so when its target rocket failed, Gemini 8 was to be the first docking in space. The mission proceeded on 16 March 1966 and was "a real smoothie" as Armstrong described the actual docking. A short circuit in

With Dave Scott (right), before Gemini 8.

one of Gemini's thruster caused it to fire, sending the spacecraft and docked Agena target into a spin, which got even faster when Armstrong separated from the target rocket. Spinning at about 70 rpm, Armstrong reported over the static, "We're spinning end over end and can't turn anything off". Eventually, using the re-entry control system, Armstrong brought Gemini under control but mission rules meant ending the flight with a premature splashdown in the Pacific Ocean, in which Armstrong and his pilot Dave Scott spent an uncomfortable three hours before being picked up.

Gemini 8 was the only mission of which both crewmembers eventually walked on the Moon. It didn't look as though Armstrong would be getting there very quickly, judging from his dead-end assignment, backing up Gemini 11 in November 1966, by which time many of his colleagues were already in the Apollo programme, training for specific flights. Indeed, it wasn t until after the Apollo 1 fire that Armstrong's name made it onto the crew rosta, backing up Apollo AS-504, which was to be the second test of the full Apollo system, in deep Earth orbit. Apollo 504 became Apollo 9, with the same objective. Armstrong became the first man on the Moon thanks to James McDivitt, the commander of AS-503. Because of production difficulties, AS-503, or Apollo 8, was not to fly with a lunar module and would, instead make the first flight to the Moon. McDivitt didn't fancy the moonflight, he wanted to see his important lunar module test through. So, Apollo 8 became Apollo 9 and Apollo 9 became Apollo 8. Armstrong, was in line for Apollo 11, instead of Apollo 12. It was as simple as that.

This crew juggle influenced many an astronaut's fate but perhaps illustrated more than ever, the fact the the first men to go to the Moon just happened to be next in the queue and were not chosen because they were the best. Armstrong honed his moon landing skills on the horrendous looking Lunar Landing Training Vehicle which resembled a flying bedstead from the 1950s. During one flight, the vehicle misbehaved and Armstrong ejected only to find himself parachuting into the now crashed and burning wreckage. The wind saved him from a nasty end.

The Apollo 11 crew were amiable strangers, thrown together to do a job. Lunar module pilot Edwin Aldrin was to be the first to step onto the Moon, according to media speculation. However, the commander's post in the lunar module was closer to the exit hatch than the LMP, so it made sense to let the commander out first. It was a classic NASA engineering decision but became an infamous one because of its historic implications. Armstrong would be first – and why not? He was, after all, the commander. There were suggestions that Armstrong made the change because he wanted to be first. Indeed, he told Aldrin "with a coolness I did not know Neil possessed," that the first steps would be historic ones and he, Armstrong, did not want to rule out, the prerogative of going out first.

The cool, quiet, enigmatic Armstrong didn't exactly set the world alight at the pre-launch press conferences. His voice was spoken in monotone, displaying no emotion or expression. The mission was a relatively quiet affair. Armstrong's last second landing on the Moon was a nail-biter, with a series of computer alarms threatening an immediate abort. Persevering to the end, he landed with just '15 to 20' seconds of fuel, according to the commander later, before an abort would have been ordered. This was naturally the high point of the mission for Armstrong. "The most exciting and rewarding part of the Apollo 11 flight was the descent to the lunar surface," he said at the press conference (not the first walk on the Moon!). After the landing, Armstrong turned to his right, smiled and shook hands with Aldrin.

It was during the wait to exit the lunar module *Eagle* that he decided what to say. The typical Armstrong would explain afterwards, that while he knew that the chances of getting back to Earth were very good, the odds of a safe moon landing were about 50:50 and why waste time working out what you're going to say until you know you're there to say them? Nobody had suggested what his first words to be. Indeed, controllers at Houston were expecting something like "There's no problem, the surface is smooth". Armstrong uttered "That's one small step for (a) man, one giant leap for mankind," and absolutely stunned Houston, particularly the public affairs people, who had never heard Neil say anything like that! Although he maintains that he said "a man", it certainly doesn't sound like he did, but as he explained, the way he talks sometimes results in one syllable words being lost particularly in crackling voice transmissions. He admitted ten years later, "as you can tell, I'm not very articulate and I drop syllables frequently. It was intended, if you would like, to put it into parentheses".[1]

There are no 'classic' pictures of Armstrong on the Moon because Aldrin didn't take any. Armstrong fixed the camera on a bracket on his chest at the start of the Moon walk and most of his pictures of Aldrin – many of them unscheduled – were taken up to the point of the flag raising ceremony which was interrupted by President Nixon, delaying the crew's work. Things got very hectic after this. Aldrin claims that his plans to photograph Armstrong were thwarted by Nixon's interruption. But at some point it was scheduled for Armstrong to put the camera down and for Aldrin to pick it up. Aldrin did pick up the camera. Among his tasks was to take a 360 degree panorama and Armstrong just happened to be in shot at the time. So Aldrin took one shot of Neil, standing in shadow and with his back to the camera. Suggestions that Aldrin did not take any photos of Armstrong because he was furious at not being the 'first' are merely media hype. Both Armstrong and Aldrin said after the flight that this period on the Moon was extremely busy (especially after Nixon's interruption). In any event, Armstrong was not the sort of person who would ask to have a 'holiday snap' taken of him. He certainly turned out to be the first and best lunar photographer.

Was there a moment when he felt spellbound on the Moon, he was asked at the press conference. "About two and half hours," he replied in deadpan manner. He proved to have a sense of humour. Asked whether there would be women astronauts, he replied, "I hope so!"

Armstrong served dutifully as a fine ambassador during the equally hectic world goodwill tour. I remember him cheerfully shaking as many hands as he could and even holding a baby whilst on the steps of the American Embassy in London. He stage-managed the press conference with an affable ease, contrasted with Aldrin's tense curtness. Goodwill tours and speeches to Congress over, astronaut Armstrong became Neil Armstrong again. He stayed with NASA at first, moving to Washington DC to became deputy associate administrator for Aeronautics. In 1971 he became a professor of engineering at the University of Cincinnati. He worked at the university until 1980, and

After the moonwalk.

hardly mentioned Apollo.

He then entered the business world, as chairman of the Cardwell International Corporation in Lebanon Kansas. In 1984, he became chairman of CTA Inc, a computing systems company. Returning to the public eye, he served on two Commissions with sadly ironic contrasts, the first looked at future space policy and having just completed that, reviewing the cause of the *Challenger* disaster, which, perhaps, made the first commission rather academic. Armstrong lives on a small grain farm near Lebanon; back as the "farm boy from Ohio," he says. He just doesn't make a thing of talking about 'it'. At his golf club, he'll not hear of it if you want to talk about the Moon but he'll talk all night about his kids and other things, reported Lebanon's golf pro, Dick James in *USA Today*[2]. Armstrong finds it hard to understand the 'recluse' tag that has been given to him. After all, as he says, he travels a good deal, is heavily involved in his own business and a number of technical and charitable organisations, makes a "couple of dozen speeches a year" and appears on TV and at public events. What could be less reclusive than that?

The reclusive executive. (CTA Inc)

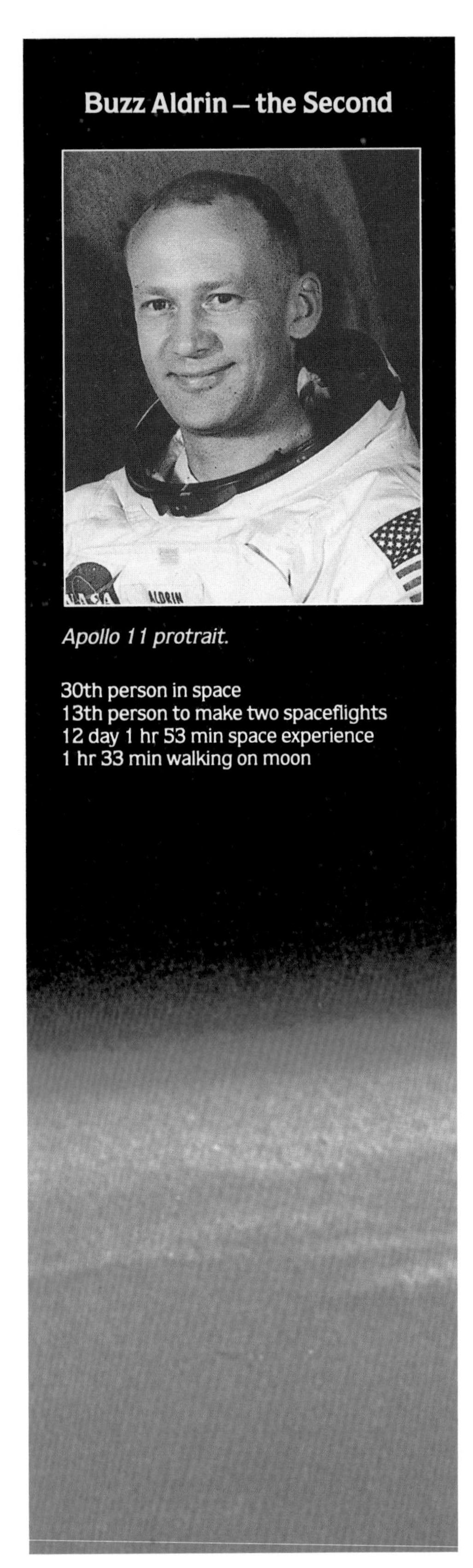

Buzz Aldrin – the Second

Apollo 11 protrait.

30th person in space
13th person to make two spaceflights
12 day 1 hr 53 min space experience
1 hr 33 min walking on moon

ACCORDING TO INCORRECT MEDIA speculation at the time, fate and the Apollo 11 checklists decreed that Buzz Aldrin was to be the first man on the Moon. But fate decreed that Aldrin was not to be commander of Apollo 11 and you couldn't blame any commander for questioning a somewhat inexplicable NASA plan that the second-in-command make the most historic steps in history. In fact it would have been a near impossibility for the LMP to get out first – because the door opened towards him. Which ever way it happened, and Aldrin says that the media got it wrong, Armstrong went out first.

But Aldrin was one of the first two men on Moon and he achieved the pinnacle of his career and a place in history. After this, however, Aldrin would ask himself, "what next?" When you've done the Moon there's nowhere else to go but down. Nothing more to achieve.

Aldrin experienced "problems that were to grow from a number of causes". As he recalled in his autobiography, *Return to Earth*, his liking for a glass a two of liqour grew to addiction. He suffered depression. He had an affair. He had difficulty making decisions. He felt tired and lethagic. So "suffering from depression and alchohol abuse I sought medical treatment", he recorded.

The post-Apollo years haven't been kind to Buzz Aldrin ; they have taken their toll on the man, whose creased and worried face is that of a seventy-year-old, not of one who is just sixty.

But all this is behind him, and a rehabilitated Aldrin is naturally keen to avoid the perpetuation of a portrait he painted of his troubled self in his book – now fifteen years old. USAF Colonel Edwin E. 'Gene' Aldrin, married to Marion Moon, was a distinguished Air Corps aviator – he still held his pilots licence at the age of 73, when his son, Buzz, took off on Apollo 11. He was also a student of Robert Goddard, the father of American rocketry and was a friend of Charles Lindbergh. Now working for Standard Oil, promoting avaition around the world, Gene Aldrin had settled his family in Montclair, New Jersey, when his son, Edwin E. Aldrin Jr, was born there on 3 January 1930. He was nicknamed Buzz by his sister, Fay Ann, and the nickname stuck, eventually making its way into the history books.

Aldrin Jr was encouraged to excel at school, especially in sports. Academically, he didn't really knuckle under until he decided that, like his father, he wanted to be a pilot. His grades at Montclair High School would have to be good if he wanted to get to West Point or another military academy. They were and he did. Along the way, Aldrin was football centre for his school team that won the New Jersey championship in 1946. He was also a 13 ft pole vaulter. Aldrin, aged 17, entered West Point and in 1951, graduated third in his class of 475.

First day at school (at NASA in October 1963), seated first left. Seated third from left is Charlie Bassett whose death in an air crash had a profound influence on Aldrin's fortune.

Aldrin entered the US Air Force and was given flight training at bases in Florida, Nevada and Texas, where he received his wings at Bray Air Force Base in 1951. He was posted to the 51st Fighter Interceptor Wing in Suwan, Korea and the evening before he departed, he met Joan Archer, of HoHoKus, New Jersey, whom he courted for three years "from a distance of thousands of miles", going out with her just four times, before taking the plunge and asking her to marry him. Aldrin flew F-86 aircraft during the Korean War and downed two enemy Migs during his 66 combat missions, earning various combat and air medals. Returning to the USA, Aldrin served as an aerial gunnery instructor at Nellis Air Force base, Nevada and attended squadron officer's school at the Air University, Maxwell Air Force Base, Alabama. In December 1955, Aldrin became aide to the dean of faculty at the new Air Force Academy. In June 1956, after getting married and fathering his first child, a boy, Michael, Aldrin became flight commander with the 36th Tactical Fighter Wing in Bitburg, West Germany. Two more children arrived in 1957 and 1958; Janice and Andrew.

In 1959, the year the Mercury astronauts were selected, Aldrin attended the Massachussetts Institute of Technology to work on a doctorate in astronautics, his thesis being guidance for manned orbital rendezvous. Aldrin applied to join the second NASA astronaut group but was turned down because he was not a test pilot. In 1963, his doctorate achieved, Aldrin worked briefly in the Gemini target office of the US Air Force Space Systems division and at NASA Houston, integrating Dept of Defence experiments into

Gemini mission planning. By this time he had applied to join the third group of NASA astronauts, since the conditions of entry had been relaxed enough to include fighter pilots. His doctorate in Gemini type rendezvous must have been a great help in his being accepted. Ironically, Buzz Aldrin didn't come out top in NASA's Gemini crew selection and wasn't even on a prime Gemini crew, despite his rendezvous knowledge, which had placed him on a NASA panel which planned Gemini manoeuvres. He was to be back-up pilot of Gemini 10, making him elegible for Gemini 13, except there wasn't a Gemini 13 mission. Then fate took Aldrin to the top of the spacewalking stakes – on a Gemini mission. Aldrin recalled his sadness that it took the death of his great friend Charlie Bassett in February 1966, to get him assigned a mission. As a result of the Gemini 9 pilot's death, with his command pilot, Elliot See, the Gemini 9 back-up crew became prime and the Gemini 10 back-ups moved up one flight, making Aldrin and his command pilot elegible for Gemini 12 in November 1966. He and Lovell walked to the launch pad with 'The' 'End' signs on their backs. During the rendezvous, Gemini's radar failed and out came Aldrin's slide rule. Rendezvous was achieved. "He's incredible," said an adoring flight controller of Aldrin.

During a record-breaking full EVA, lasting over two hours, Aldrin at last laid to rest the doubts about man's ability to work in space, which had been raised by the troubled EVAs of past missions. His was a cool, clinical mission which ended a run of ten Gemini missions. Apollo beckoned – but which one? Having been assigned to the last Gemini, Aldrin was lower down on the rungs of the Apollo ladder, making his first appearance as back-up lunar module pilot of AS-504 which was to be a test of the lunar module in deep Earth orbit. His commander was Neil Armstrong and command module pilot Jame Lovell. But being on the lower rungs early meant that Aldrin was higher up when it mattered. An early flight to the Moon was

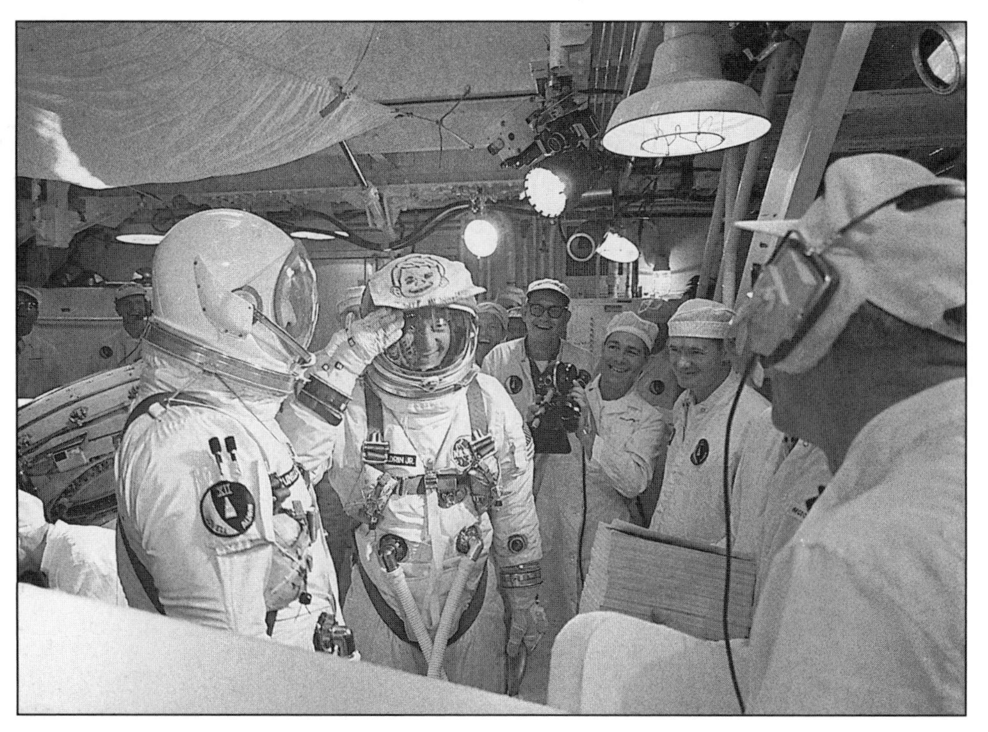

Jokes before lift-off on Gemini 12.

clearly in the offing and Aldrin, the LMP, was not just going there, he was going to walk on it too. Then AS-504 became Apollo 8 in 1968 and headed for Moon orbit without a lunar module. To complicate matters, prime command module pilot for Apollo 8, Mike Collins, had to drop himself from the mission to have an operation on his spine and his place was taken by Lovell, who was then replaced by Aldrin, leaving the LMP spot for new man Fred Haise. Aldrin was a CMP and if this crew was rotated as Apollo 11, he wouldn't be landing on the Moon. By January 1969, Collins had recovered and was placed back into the first available crew, Apollo 11, with Aldrin back as LMP, replacing Fred Haise, who as a rookie was unlikely to have made the mission anyway.

Armstrong, Aldrin and Collins were men apart. They rarely socialised and just got on with the job. According to several reports, the flight plan had Aldrin going out of the LM first on what is now clearly going to be the first landing attempt. Aldrin says that there was no firm plan and that a decision was being left until "late in the project". The rumours started flying. Armstrong would be first because he was civilian and it would be better for a NASA employee to take the first steps, rather than a military officer. Armstrong said nothing about it. Eventually Aldrin, who felt that rumours were distracting and that the uncertainty was hurting mission planning, tackled the enigmatic Armstrong who replied that making the first steps would be historic and he didn't want to rule out the prerogative of going first. It was out at last. Armstrong would be first. Was it Armstrong who pulled rank and demanded the change or was it the NASA big wigs? Armstrong says he never knew about the decision, until it was made: the matter had been settled "one way or another" and it was a entirely logical one given the logistics.

The question has been asked, however, why it was that Aldrin took no photographs of Armstrong. During simulations of the moon walks, the astronauts exchanged the camera, which public affairs people had hoped would be used at some time to take pictures of each of them, particularly Armstrong, the first man on the Moon. Amazingly, 'photos for the our public affairs friends' were always taken reluctantly by the astronauts and Apollo 11 scientists had not even considered taking colour film to the Moon. A black and white picture of the first men on the Moon on the front cover of Life magazine? The scientists relented and some 'tourist' photography was scheduled, including some in colour. However, due to recent changes in the management of the public affairs department at Houston after the departure of the 'voice of Gemini' Paul Haney, and to the main priorities of the Apollo 11 mission, public affairs photography was never high on the agenda. The flight plan showed Armstrong, who, incidentally did the PR guys proud with superb pictures, putting the camera on a pedestal and leaving it for Aldrin to pick up. The flight plan also instructed Armstrong to take those famous pictures of Aldrin but, in fact, not as many as he actually took. Probably, being the sort of chap he is, Armstrong would never say, "Oh, Buzz take a few shots of me, won't you". This was very un-Right Stuff. Armstrong today is quizzically amused about any fuss being made about this episode. Buzz, as instructed in the flight plan took some pictures of his bootprints in the lunar dust. Buzz did take one picture of Armstrong when his commander was in shadow and had his back to the camera! This resulted because Aldrin, following the flight plan, was taking a 360 degree panorama for documentary purposes and Armstrong just happened to be in shot at the time. The flight plan did not instruct Aldrin to

The epic photo of Aldrin taken by Armstrong, reflected in Aldrin's visor.

'take a a photo of the commander'. Recalling the chaos over pre-mission planning for the moonwalk, one public affairs official felt that it was such a period of intense activity, that Aldrin didn't even get a chance to think about it and if Armstrong had asked him, then he would have taken a photo. Judging from the comments both made about their period on the surface, he was probably right. In any event, even if Buzz had been a bit sore (he insists that he wasn't) he was as straight as a die.

Aldrin had been trained for Apollo but not for his post-Apollo mission. Within days of being let out of the quarantine container, the shy, reticent loner was making a speech to the full Congress of the United States. He was being bundled around the world on the goodwill tour that stretched the goodwill even of the affable Mike Collins. The schedule was quite ridiculous: Mexico, 29 September, Columbia, 30 September, Argentine, 1 October, Brazil, 2 October . . . Guam, 2 November, Korea, 3 November, Japan, 4 November . . . then home. Whew! Aldrin looked morose and irritable when I saw him at a press conference in the US Embassy in London; he was having a rough splashdown. What on Earth was he going to do now? His highly structured life always had a goal. What could that goal be after being to the Moon? "Without a goal I was like an inert ping pong ball being batted about by my whims and the motivations of others. I was suffering from what poets have described as the melancholy of all things done," he would write later in his autobiography.

For a while, he continued at NASA, working on preliminary designs of the Shuttle and hosting a tour by two Soviet cosmonauts but jacked it in "because of doubts about certain aspects of the shuttle programme", to take on the very prestigious post of commander of the test pilot school at Edwards Air Force Base. Depression and alcoholism took their toll and Aldrin sought medical assistance. After being hospitalised for a month, at Brooks Air Force base, he tried to make a clean break of things, resigning from the Air Force and, after much deliberation, divorcing Joan. In his book, *Return to Earth*, he described his feelings in detail in a moving and candid way. His story was told in a TV film starring Cliff Robertson.

Ten years after. The heavy burden of life after the Moon is showing.

Nearly 20 years on. (John Hughes)

Aldrin became honorary chairman of the National Association for Mental Health and attempted various business ventures which never quite took off. He even did some TV car ads and appeared in a TV movie. Aldrin got married again. But life wasn't easy. He had setbacks. His second marriage failed. In 1985, Aldrin became professor at the Centre for Science at the University of Dakota in Grand Forks.

In 1988 he was married again, to Lois on Valentines's Day, and when I met him at the Farnborough Air Show, was living in California and working as a busy space consultant and also writing a book about the moon flight. After reading *Return to Earth* one might be forgiven for thinking that every time Buzz Aldrin looks up at the Moon, he probably doesn't recall in awe the wonders of his hours up there. He probably just waves his fist at it." Nonsense," says Aldrin, who is proud of all that he and his fellow astronauts did.

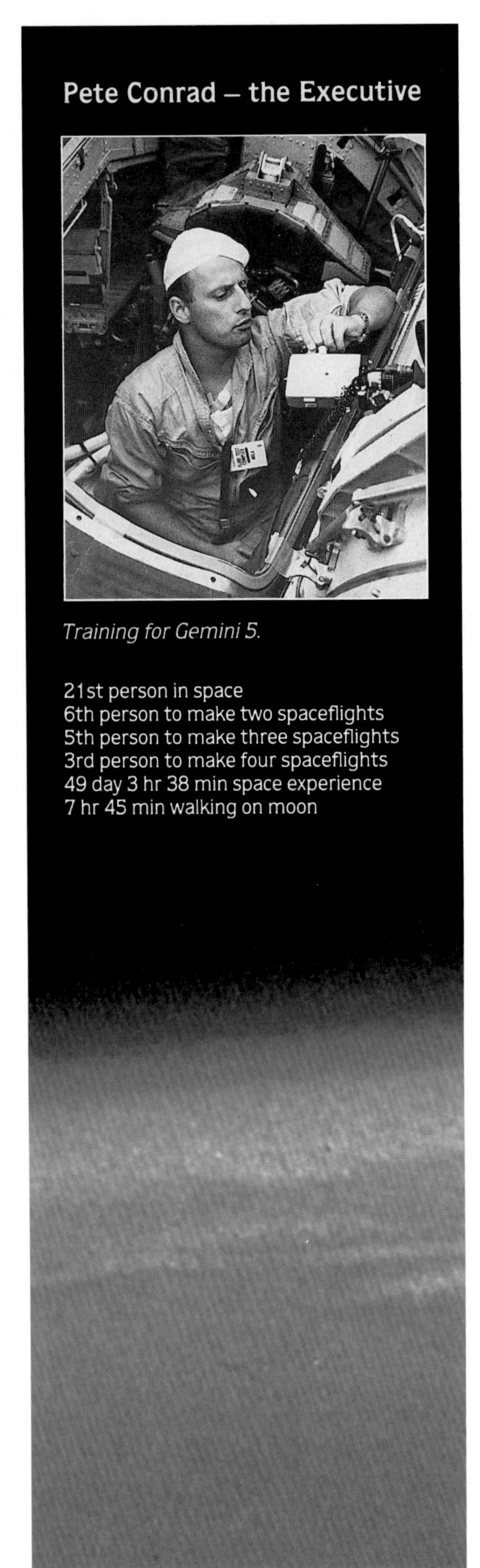

Pete Conrad – the Executive

Training for Gemini 5.

21st person in space
6th person to make two spaceflights
5th person to make three spaceflights
3rd person to make four spaceflights
49 day 3 hr 38 min space experience
7 hr 45 min walking on moon

PETE CONRAD HAD NO ILLUSIONS ABOUT his fame. He was the third man on the Moon. "Who remembers the third man to fly the Atlantic?" he often asks, in reply to questions about what effect being the third man on the Moon had on his life. That may be, but would he have been snapped up by McDonnell Douglas to help market its new DC-10, if he hadn't been to the Moon?

The two sides to Conrad's post-Moon story combined vividly for me at the Paris Air Show in 1987. Conrad had been moved around in the executive levels of McDonnell Douglas for some years and was at the time vice president of international business. Just turned 57, the diminutive businessman, with bald head, thin grey hair and, of course, the famous gap-tooth – together with remarkable twirling eyebrows which swoop over his eyes like Superman curls, giving the appearence of an owl – was sheltering from the seemingly perpetual rain that showered the Show that year, wearing a plastic raincoat and standing on the porch of the McDonnell Douglas chalet, waiting for a car to pick him up. I went up to him and said "hello". He shook my hand and immediately said, "Hi, Pete Conrad," in those characteristic gravelly tones. Americans always seem tell you who they are when you have addressed them. Perhaps, Conrad had spent so long as a fairly anonymous Moonman, that he felt compelled to tell people who he was. Anyway, we had a chat and the avid motor racing fan said he was looking forward to the Le Mans race and thought the weather pretty poor for it. We said goodbye. I went inside the chalet. Soon, the round-the-world-without-stopping pilots, Rutan and Yeager came into the reception, followed by a bustle of photographers, to have their picture taken with some McDonnell Douglas guests. The glass doors were flung open as the entourage spilled out of the chalet and poor Conrad was pushed to one side, stuck behind the opened glass door. The cameras flashed away at the latest heroes, while a hero of the past was plastered against a wall. I recalled Conrad's words. "Who remembers the third man to fly the Atlantic?"

Charles Conrad Jr was born on 2 June 1930 in Philadelphia, Pennsylvania. He was the son of Charles Conrad Sr and the former Frances Vinson, who wanted to call Conrad Jr, Peter. Conrad Sr wanted to call Conrad Jr, Charles but Frances didn't want Charles Jr to be called Junior, so Charles Jr became known as Pete. He attended primary and secondary schools in Haverford, Pennsylvania and in New Lebanon, New York. At the age of 14, Pete Conrad walked into the local airport at Paoli, Pennsylvania and told the manager, "I

With Cooper (right), before Gemini 5.

With Gordon (left), after Gemini 11.

The third man on the Moon.

want to learn to fly. I want to learn all about aeroplane motors".

He attended Princeton University and graduated from there in 1953 with a science degree. The following day, he married Jane BuBose of Uvalde, Texas. While at Princeton, Conrad had been a member of the Naval Reserve Officers Training Corps Unit and on graduating, was commissioned in the Navy as an ensign, on 5 June 1953 at the Naval Air Station at Pensacola, Florida. He continued flight training at Corpus Christi in Texas and on 24 September 1954 became a Naval aviator, joining Fighter Squadron 43 the following month.

His mother was no doubt pleased that Pete and Jane's first child, a son born in December 1955, was christened Peter, now a rather portly, moustached and slightly balding 34-year-old. Another son, Thomas, arrived in May 1957.

The following March, Conrad made his big step towards an illustrious space career, hardly matched by any other astronaut, by moving to Pax River to train as a test pilot. The Naval Air Test Centre at Patuxent River, Maryland was his home until February 1961 by which time he had fathered two more boys, Andrew, born in April 1959, and Christopher, born in November 1960. After graduating as a test pilot, Conrad had been assigned as a test pilot at the flight centre's armanents division, then as an instructor and finally a performance engineer. His name was put forward as a potential Mercury astronaut and after a range of wierd and wonderful, and sometimes horrific, medical and psychological tests, Conrad came close to being chosen. His name was well known at NASA for the second round of selections. In 19 March61, a month before the Mercury Seven were named, Conrad was assigned first to Fighter Squadron 121 until October 1961, then to Fighter Squadron 96. By October 1962, Conrad was reporting for duty at the NASA's Manned Space Flight Centre, as one of the nine second division astronauts. At just 5 ft 6½ inches and weighing 138 pounds, Conrad was also the smallest.

Conrad was the third of the team to be chosen for a flight, after John Young and Ed White. He was assigned to pilot Gemini 5 with Gordon Cooper. The mission was to be the one in which America would at last take the lead in the manned spaceflight duration stakes. Scheduled to last eight days, Gemini blasted off on time on a Saturday afternoon, 21 August 1965. The spacecraft contained a radar evaluation pod which was to be released and with which Gemini would rendezvous later, during the first exercise of its kind ever performed in space. Another unique feature of the flight was that Gemini 5 included fuel cells to produce electricity and it was these that failed almost completely, threatening to end the mission after just a few orbits. With a powered-down spacecraft, the space duo sat in a confined space equivalent to the front seat of a car with nothing much to do than look out of the window. Conrad later recalled that because he and Cooper had trained together for so long, they had swopped all their sea shanties and stories and didn't have much to say each other. With little physical activity, the astronauts found difficulty sleeping and even though the view out of the window was great, even that became monotonous after eight days.

Teamed with Dick Gordon, Conrad backed up Gemini 8 and took Gemini 11 to the skies in September 1966. This flight was by no means boring, making a simulated take-off from the Moon and rendezvousing with a target rocket in one orbit, as would be performed during Apollo. Using the Agena target's engine, Gemini 11 was propelled to a record height of 875 miles over Australia. "The world is round!" shouted Pete. A tether was attached to the Agena during a curtailed spacewalk by Gordon and when it separated from Gemini, was used to create artificial gravity in space. Finally, for the retro fire and splashdown, Conrad and Gordon folded their arms and it all happened automatically under computer control, with Gemini 11 almost hitting a bulls-eye target. It was a terrific mission and as usual, the grinning, gapped teeth were splashed all over the newspapers. The elated Conrad moved to the back-up crew of Apollo 3 with Gordon and C. C. Williams.

The usual juggle of missions followed. Apollo 3, which was to have been a high-altitude test of the lunar module, eventually became Apollo 504, by which time Williams had been killed in an air crash and replaced by Al Bean, making the team – all friends – an all-Navy affair. If all had gone according to plan, AS-504 would have become Apollo 8 and Conrad would have taken Apollo 11 to the Moon. Instead, because Jim McDivitt, the commander of 504, refused to relinquish his mission, even if it meant his missing the first lunar orbital flight, it became Apollo 9 and Conrad became the third man on the Moon instead of the first.

Apollo 12 was an exciting flight for Conrad – and for all who listended in to his amusing banter with Bean – but as for going to the Moon, it just seemed a logical thing to do having spent seven years sleeping, eating and breathing the Moon. Apollo 12 didn't change Conrad. He was the same exhuberant and amusing character after it. Apollo 12 was not even his finest hour, according to the astronaut. That was Skylab, a project he sort of highjacked after the cancellation of Apollo 20 to which he had been unofficially assigned to command, to become the first man to make two Moon landings. The unique mission was lost by budget cuts. He almost didn't make Skylab. On 11 May 1972 his T-38 jet ploughed into a field near Bergstrom, Texas, only the tail plane being left recognisable in the wreck in which, fortunately, Conrad was not lying because he had parachuted to safety. He had been to Dover, Delaware for a spacesuit fitting and ran out of fuel after being diverted twice from his original destination, Ellington Air Force Base.

Without a doubt, Skylab was, indeed, Conrad's finest hour – or rather 28 days – during which he saved the entire programme with a dangerous, unscheduled spacewalk which, as the years go by, takes on even more heroic proportions. Skylab was America's first space station born of the Apollo Applications programme, which used left-over Apollo hardware. It basically consisted of an empty S4B Saturn 5 third stage, equipped to house three astronauts who would conduct 270 life sciences, solar physics, Earth observation, astrophysics, materials processing, engineering and technology experiments, during three long duration missions lasting 28 days. The duration eventually became 59 days and 84 days respectively, after extensions of the original plan. Skylab was all America had left for its astronauts until the Shuttle came along, except the US-Soviet jamboree in 1975. A lot depended on a successful Skylab programme to keep America in front in space.

Skylab was lifted off pad 39A on a cloudy morning at the Kennedy Space Centre on 14 May 1973 by the final Saturn 5 booster to grace the air, or bludgeon its way through it, depending how close you were to the blast off. Observers weren't to see a catastrophic series of events to occur 65

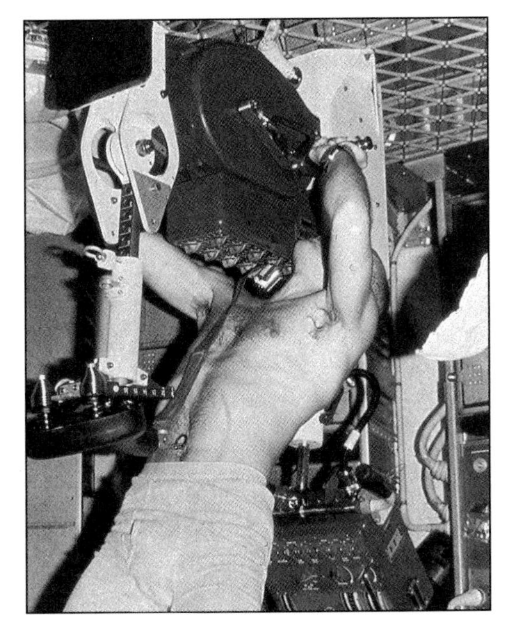

Aboard Skylab.

Nearly 20 years after the Moon. (McDonnell Douglas)

seconds later. Flying through thick cloud, Skylab's micrometeorite and thermal protection shield tore lose. This in turn took with it one of the space station's solar panels. Once orbit had been achieved, it became clear that the second panel had jammed. Skylab was a useless multi-million pound space station, and the largest object in space to boot. Conrad and his two crewmen, Navy medic Joe Kerwin and Navy pal Paul Weitz, who would have flown as CMP to Conrad on Apollo 20, were due to have taken off the following day. Their mission was delayed to 25 May during which period the crew practiced procedures to place a new shield over Skylab and pull the solar panel out.

Skylab 2, as the Apollo capsule was called, lifted off on a Saturn 1B and soon Conrad was shouting "Tally Ho! Skylab," and confirming the worst suspicions about the state of the station. Flying in close, Conrad steadied Skylab 2 while Weitz opened the hatch and, with Kerwin clinging onto his ankles for dear life, tried to pull the panel out using a hook on the end of a pole. It was useless. So, it seemed, would be the docking. Seven times they tried, before success. Once inside, the crew restored temperatures to reasonable levels by deploying an umbrella-like sunshield over the capsule. But power shortage was critical. Without that panel, the other missions would have to be cancelled. So, on the fourteenth day in Skylab, Conrad risked his life by going EVA with Kerwin. Using wire cutters, he freed the panel of its obstructions and physically pulled it out. As the panel sprung out, Conrad was nearly bundled to his doom. There have been brave EVAs since Conrad and Kerwin's but their's was the first and bravest.

Conrad brought his crew home after 28 days in space, his second manned duration record and 20 days longer than his first. Skylab housed two more crews, allowing the last to set a record of 84 days which today palls into insignificance compared with year-long flights by the Soviets. More because of this feat than the Moonwalk, Conrad was awarded the Congressional Space Medal of Honour by a distinctly anti-space President Carter. Conrad was also elected to the Aviation Hall of Fame.

In December 1973, Conrad left Nasa and the US Navy to pursue a business career. First he became vice president of the ATC TV company in Denver, Colorado. In March 1976, Conrad stepped onto a middle rung of the executive ladder of McDonnell Douglas, becoming a vice president and consultant to the mammoth aerospace company, which Conrad had been supporting by being featured in ads promoting the DC10, an aircraft which was earning itself some bad PR. The name Conrad was still widely recalled in the mid-1970s and Conrad retained his celebrity status. But as the years went by and as he was moved around the McDonnell Douglas hierarchy, it was his business performance and not Moonman status that became more important. Conrad, as ever the enthusiastic, entertaining and exhuberant man, became vice president of commercial sales for Douglas Aircraft, vice president for marketing and senior vice president marketing. In 19 February83, he was made senior vice president for marketing and product support. A year later he was promoted to corporate vice president in that confusing reporting structure that Americans seem to love. He then became vice president for space programme development, even making EVA simulations in a water tank to support his company's development work for the Space Station. In 1986, he was made staff vice president for international business development, on duty at the Paris Air Show in 1987. Asked at the Farnborough Air Show in 1988, whether he was writing a book about his experiences, like some former fellow astronauts, Conrad said, "if I could write, I wouldn't have become an astronaut".

Alan Bean – the Painter

Last in the Class 3 astronauts to fly, Bean in 1963.

45th person in space
22nd person to make two spaceflights
69 day 15 hr 45 min space experience
6 hr 50 min walking on moon

WHEN ALL THE MOONWALKERS ARE DEAD and buried, the chances are that the one whose name will live on more than the first, will be the fourth. Yet Alan LaVern Bean is not known as an astronaut anymore. He is an artist and a space artist to boot. His impressions of the Moon, put to canvas in his small condomonium-cum-studio in Texas, "2.7 per cent smaller than Skylab,"[1] are already being compared, in quality and style, with the work of famous American artists, such as Norman Rockwell and space artist Robert McCall. The pictures will be Bean's legacy to the people who didn't go to the Moon, his expression of the ultimate experience. The irony is that Bean, more than any other of the moonwalkers, owes his place on the Moon to the death of a colleague. Bean became the fourth man on the Moon on 5 October 1967 when astronaut Clifton C. 'C. C.' Williams's T-38 ploughed into the ground near Tallahassee, Florida. Williams was on the Conrad crew and Bean was working on the Skylab space station project, without any thoughts of making it to the Moon.

Bean was born in Wheeler, Texas on 15 March 1932 and grew up in Fort Worth where he graduated from Paschal High School in 1949. Bean wanted to fly. He wanted to join the Navy. Navy pilots are the best because they have to land on carriers, he thought. His first steps in that direction were thwarted somewhat by his mother, Frances, who refused to sign his papers for him to enlist in the US Navy Reserve. His father, Arnold, duly obliged and on 4 May 1949 he became an electronics technician striker at the Naval Air Station in Dallas. He was honourably discharged on 20 September the following year when he entered the University of Texas to gain a science degree in aeronautical engineering. He didn't want to be an engineer but a better pilot, he said afterwards.

While studying, Bean was a member of the Reserve Officer's Training Corps and while in the gym met Susan Ragsdale of Dallas. The couple married in 1955, the year Bean graduated and became Ensign Bean on 29 January. Their first child, Clay, was born the following December.

Bean was trained to fly at the Naval Air Station in Pensacola, Florida and, from February 1956, at the Naval Auxiliary Air Station, Chase Field, Beeville, Texas. After receiving his wings on 6 June 1956 he was assigned to Attack Squadron 44, based at Jacksonville, Florida until February 1960, when the now Lt Bean went to the US Navy's Test Pilot School at Patuxent River, Maryland. He offically became a test pilot in November 1960 and was assigned to remain at Patuxent as Attack Plane Project Officer, flying every type of aircraft, from jets to helicopters, including the first tests of the A5A and A4A attack planes. In February 1963, a month after his daughter Amy Sue was born, Bean moved to Los Angeles and the University of Southern California's School of Aviation Safety. Upon completing the course of training he was assigned to the Attack Squadron 44 in May, moving to Attack Squadron 172 at Cecil Field air base in Florida, to fly operational A4 jets as Safety Officer, in October 1963. This was rather academic as it turned out because the same month, Bean was named as one of the 14 new astronauts selected by NASA.

He had been persuaded to apply by his friend and mentor, Pete Conrad, who had made the second group after failing to make the Mercury Seven eight. Bean, who made the finals for the second group and failed, went to Houston in January 1964 and reported for duty and a space career lasting 15 years.

Bean's initial space career was an extremely frustrating one. After the first Group 3 astronaut, Dave Scott, had been assigned to Gemini 8 in 1965, Bean hoped for Gemini 9, then 10, then 11 or maybe 12. The call never came. Not, that is, until Gemini 9's original crew, Elliott See and Charlie Bassett, were killed in particularly tragic circumstances. Flying their T-38 jet to the McDonnell Douglas factory at St Louis to check out their Gemini 9 capsule, the pilots encountered thick fog and waved off the first landing attempt. On the second, the jet hit the roof of the very building which housed Gemini, bounced off then hit a car park, exploding instantly. Bean was drafted into Gemini as back-up command pilot of mission 10, with back-up pilot C.C. Williams. This was a dead-end job because there would be no Gemini 13. Williams was assigned to work with Conrad and Gordon on a future Apollo mission, while Bean was assigned to manage Apollo Applications programme which would later become Skylab. It was a post he wanted because it would mean commanding the first mission and he wanted to do something 'that was first'[3]. Then 'CC' was killed, changing the gameplan.

Standing (right), with Gordon (centre), and Conrad, before Apollo 12.

Nineteen sixty-seven was the year of the Apollo 1 fire, which killed Grissom, White and Chaffee, and one in which four other astronauts would die: X-15 pilot Mike Adams, USAF Manned Orbital Laboratory astronaut Robert Lawrence, the first negro to be selected, and NASA's Ed Givens and 'CC'. An eighth, Soviet cosmonaut Vladimir Komarov became the first in-flight fatality. A kit bag left in the rear seat of the jet flying 'CC' home to Houston from the Cape, to be with his wife after she told him she was expecting their first child, jammed the controls. After

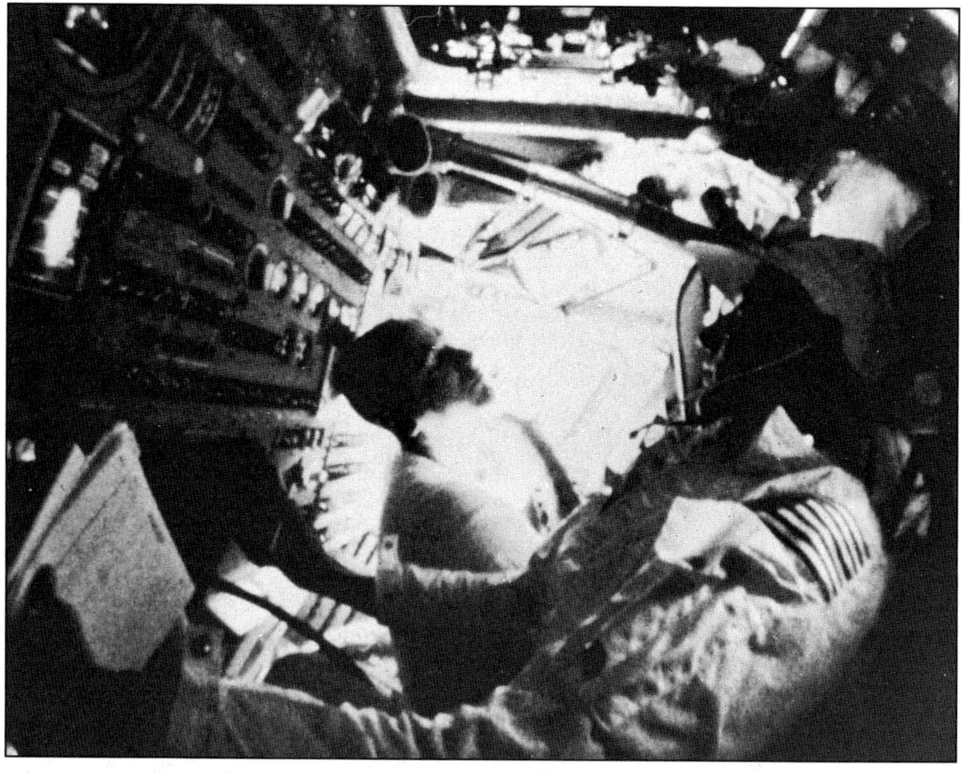

Asleep aboard Yankee Clipper.

burying 'CC,' Conrad asked Bean to join him on Apollo AS-504, which eventually became Apollo 12. Bean was the last of the Group 3 astronauts still alive to make a flight and sometimes felt that this reflected on his ability. If he needed any reassurance, he became the youngest captain in the US Navy upon his return from Apollo 12. Bean immediately went back to the Skylab project, to serve as deputy to Walter Cunningham, the likely first commander. When Conrad took over Skylab after his Apollo 20 mission was shot down in flames by President Nixon, Bean was nominated to command the second mission and poor Cunningham merely to back-up the third.

At this stage, Bean's experience of the Moon was yet to manifest itself. He was a test pilot and astronaut and getting on with it. There were no philosophical feelings about the Moon, he said afterwards. The feelings were more about surviving and the technical aspects. Anytime you let your mind wander, you are not doing the job. In high activity phases, you just can't do the job if you let that happen, he said. "One of the skills you learn as a test pilot is to keep your mind on the the job and that's the skill you develop and the one you have to call on frequently in this business. You couldn't daydream much, you didn't want any new ideas, it was hectic enough as it was. A new thought meant an old one had to be pushed aside, and you needed all the facts so you wouldn't forget what to do. I remember looking up at the Earth and saying to myself 'how beautiful', then I had to get back to work." Bean was more concerned about the effect the moon flight was having on his son Clay who was suffering the "how do I follow that?" feeling. The former Sunday school teacher, Bean, said that he experienced no religious feelings about going to the Moon.

Bean's second spaceflight began atop a

Aboard Skylab.

Saturn 1B rocket at Pad 39B, sitting on a pedestal to make it as long as a Saturn 5. His record-breaking 59 day orbital mission with science pilot Owen Garriott and pilot Jack Lousma was handled superbly. The totally unruffled crew, politely and cheerfully took to their tasks and endeared themselves highly to Houston, which is something that can't be said of the next Skylab crew. There was a scare during Bean's mission. The docked command/service module, in which the crew was to return to Earth, leaked propellant, which meant that unchecked, the spacecraft would be unreliable and the crew would require a rescue mission, commanded by Vance Brand and pilot by Don Lind, to bring home five people. The unruffled Bean's reaction to the crisis was that he thought it 'unusual'. The problem was rectified and Brand and Lind's unusual flight was not needed and Lind had to wait another 12 years to get into space on the Shuttle.

Bean tested a prototype of a manned manoeuvring unit inside Skylab and made a spacewalk outside it. The crew came home happy and fit, thanks to a regular exercise routine to stave off the ill effects of weightlessness. To Bean, Skylab, not Apollo 12, was the highlight of his space career. It was his 'Super Bowl' he said, in his interview with *Starlog*. His finest hour. It was not just a trip as such but a job lasting 59 days and one in which he and his crew made an exceptional effort, achieving 150 per cent success according to pre-mission aims. It was after Skylab that Bean began to feel the Moon effect, not in terms of his career but his personality. Later he would say, "I became overtly what I had been covertly"[2]. A lot of personal characteristics had to be concealed to get the astronaut job, to get assigned, and were concealed by the sheer intensity of the job itself. The person the astronaut wanted to project or the person the job projected became instead, the real person. "It is like being in the front line. You are risking your life and after you get through it you tell yourself you are going to live life differently now," he said[3].

For his next mission, Bean had to learn to speak Russian for he was the back-up commander of the Apollo part of the space jamboree all-in-the-cause-of-detente ASTP mission in July 1975. He trained with Ron Evans and Jack Lousma in support of Tom Stafford, Vance Brand and old Deke Slayton who got a flight at last, at the age of 51. After this mission the career astronaut started out on the Space Shuttle, his marriage having ended. In 1978, when the chief astronaut assigned himself to the first Shuttle mission, Bean, the second ranking astronaut in seniority, took over and supervised the training of the 35 new astronaut candidates. He was widely tipped for an early Shuttle mission, possibly the sixth, carrying the Spacelab laboratory.

His decision to resign from NASA in June 1981, having seen the Shuttle make its maiden flight, surprised many, and to seek fame and fortune as a professional artist seemed amazing, except to his artist friends who knew his expertise. Bean didn't suddenly become an artist, he had been painting for years – painting was listed as one of his hobbies in the Apollo 12 press kit – and during the ASTP mission he forged a friendship with fellow artist, cosmonaut Alexei Leonov. "He's just the nicest possible guy," Bean said. Taking two weeks leave and just painting, he was trying to simulate the life of an artist. He liked it but thought that maybe he liked it as a hobby but would not if it was job as described in his interview with *Air and Space* magazine. The more he painted, the more he liked it and found he actually needed to do it. "There are people who can fly the Shuttle better – or worse – than me, but I'm the only artist who'e been to the Moon, I know the stories, the hardware. So the decision was that simple," he said.[4] He figured that if he didn't make money, he'd work in Jack in the Box, a fast food café nearby.

Strangely he hadn't been painting the Moon but flowers and fruit. This went on for years; he wanted to be an impressionist painter like Monet. The Moon was grey, the sky black and spacesuits were white. Bean liked colour. One day in 1974, he started a lunar painting. Hours of total immersion later, Bean suddenly realised that this was what he should be doing. "I knew only the basics about flowers and lilies but when I paint a spacesuit, I know everything about it"[5]. Bean's scrupulously accurate as well as excellent paintings caught the admiration of fellow artists and fellow moonwalkers would take a look, reminding Bean of this and that. Bean also knew a lot of the inside stories of the Moonwalks and could call on these incidents to paint as well as the famous moments such as Scott's Feather at Hadley. There were moments to capture such as the

Bean in 1987.

time Cernan came upon a dust covered rock and, taking a sample, scrawled the name of his daughter across the dust. Bean recreated this scene as 'Tracy's Rock'. Bean has painted over 50 moon scenes and has sold many at about $12,000 a picture. No other moonwalker has bought one, however. Bean does'nt want to sell all of them, he has his favourites, such as one of Scott on the Moon. The number of moon scapes in his brain are limitless and eventually he may graduate to other scenes based on his experiences in space and others aboard Skylab and Shuttle.

Given the chance to fly the Shuttle as an artist he would turn it down. "I've been there, I've experienced it. Someone else should have the chance." He's as happy as a sandboy, living with his second wife Leslie, who calls him 'Beano', his family of five Lhasa dogs, his collection of 45s constantly playing as he works. Ivy nurtured from cuttings taken from Monet's garden and Van Gough's grave, profilerates in his small garden. "Nobody hassles me," he told *Air and Space* magazine.

Bean is communicating and expressing his Moon experience on canvas. In 20 or 25 years all the Apollo people will be gone. "I have a sense of urgency about producing good paintings and telling the story, because my time is limited too."[6] He wants to paint the history of the space age as it really was by someone who saw it and actually lived it. "I want people who wonder about it in 20 years time to say 'let's go and look at an Alan Bean painting, that's how it really was'". Bean may have walked on the Ocean of Storms but when the National Gallery of Art hangs one of his paintings[7], he really will feel he has arrived, if he's still around to see it.

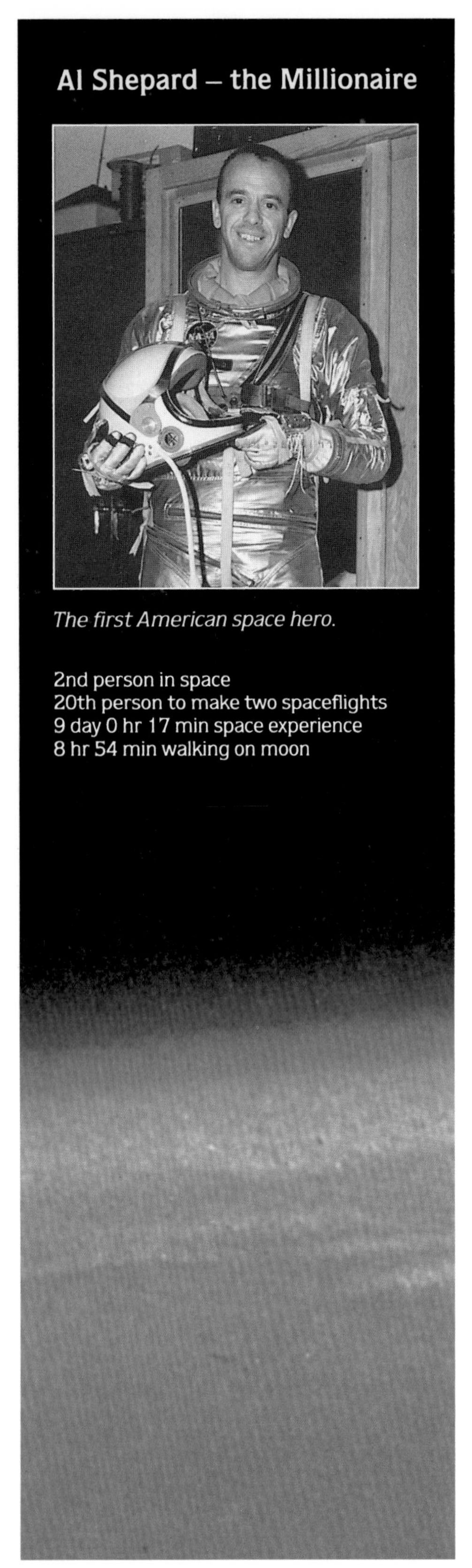

Al Shepard – the Millionaire

The first American space hero.

2nd person in space
20th person to make two spaceflights
9 day 0 hr 17 min space experience
8 hr 54 min walking on moon

CDR ALAN B. SHEPARD JR WAS AMERICA'S first man in space. But he was an American space hero for just 291 days. Shepard went up and down on a suborbital lob. John Glenn became the first to orbit. Glenn soon became known as America's 'real' first man in space. This rankled the brash and cocky naval aviator, particularly because Glenn wasn't exactly his cup of tea. Shepard heckled for an orbital flight, the last one, lasting two days, but Nasa said no. Gemini beckoned and Shepard went into training for the first mission. He was then felled by an illness which confined him to an office desk for five years, leading the astronaut corps with his feet firmly on the ground. During this period he made shrewd business connections and investments and was a millionaire by the time he recovered from his condition.

Almost immediately he was named to command Apollo 14, causing consternation in the astronaut corps which he had led with a firmness that reduced many a recruit to wobbly jelly in his presence. His very stare could cut a man dead at twenty paces, as I had the misfortune to witness in 1971. Gatecrashing an astronaut party at the Holiday Inn at Cocoa Beach before Apollo 15, with radio space reporter Dave Denault, I met Shepard, whose picture used to hang on my bedroom wall at home. My boyhood hero. I shook his hand and he gave me the Shepard stare. No words came from his frozen mouth. Denault asked Shepard what it was like to be back as chief of the astronauts after his Apollo 14 mission. A pretty harmless question. Like the chief witch in Roald Dahl's book *The Witches*, Shepard's stare reached the intensity of a laser and Denault was reduced to ashes on the floor. I retreated rapidly. Al wasn't the most popular guy around; it was his very strong, single-minded personality that got him to the Moon. And he hasn't done badly since either. A retired Rear Admiral, Shepard is chairman, president and director of several companies and banks and is commissioner for the Port of Houston.

Shepard's maritime connections were forged despite his father being a West Point graduate and US Army servicemen. Alan Bartlett Shepard was born in East Derry, New Hampshire, on 18 November 1923 and attended school there and at the nearby town of Derry. Shepard went to the Admiral Farragut Academy in New Jersey and in 1944, graduated from Naval Academy, with a science degree and from Annapolis served on the destroyer *USS Cogswell* in the Pacific during the end of the Second World War.

Impatient to learn to fly, Shepard paid for private lessons and had a private pilots licence by the time he entered naval flight training at Corpus Christi, Texas and Pensacola, Florida. He got his Navy wings in 1947, married Louise Brewer of Kennett Sq, Pennsylvania, becoming the father of Laura, and was assigned to Fighter Squadron 42 at Norfolk, Virginia and in Jacksonville, Florida, also serving several tours of duty on carriers in the Mediterranean.

By 1950, Shepard had moved to Patuxent River, Maryland for Naval test pilot training. After graduation, Shepard flew the Banshee F2H3 in in-flight refuelling tests, conducted carrier suitability tests and made the first landings on the first angled carrier deck. He flew Banshees on night fighter duty with Fighter Squadron 193 at Moffett Field, California and, as operations officer of this squadron, made two tours of duty in the western Pacific aboard the *USS Oriskany*. Returning to Patuxent for a second tour of

Lift-off on 5 May 1961.

duty, Shepard, by this time the father of Julia, born in 1951, flight tested the F3H Demon, F8U Crusader, F4D Skyray and the F11F Tigercat. He was also project officer for the F5D Skylancer. His final months at Patuxent were spent as an instructor. Shepard was then assigned Naval War College at Newport, Rhode Island from where he graduated in 1957, being assigned to the staff of the Commander in Chief, Atlantic Fleet as an aircraft readiness officer, preparing to take command of his own carrier squadron.

In the film *The Right Stuff*, based on Tom Wolfe's book of the same title, Shepard is pictured after landing on the deck of a carrier being approached by two sea-sick government officials who ask him whether he is interested in joining the space programme: "Sounds dangerous, count me in," he says with a broad grin across his rubbery face. Shepard said afterwards that much of

The grounded, frustrated Shepard follows the Gemini 11 crew.

Off again, at last, Shepard takes a final look at his Saturn 5.

The fifth man on the Moon.

the film was pure fiction and some was fact. Although typical Shepard, that carrier scene was fiction. What wasn't, was Shepard's dominance of the legendary Mercury Seven astronauts and his dual personality, 'Smilin' Al' or 'Laser Eye'. Shepard could be an absolutely charmer and loved jokes, particularly comedian Bill Dana's portrait of a Mexican astronaut Jose Jiminez. Asked whether what he was holding was a helmet, the nervous Jiminez says, "Oh, I hope not!" Shepard loved to mimick this. But to those he didn't have time for for, he could be overwhelmingly difficult.

The astronaut group split into two factions, Shepard and Glenn. The Boy Scout, good-boy Glenn was clearly impressing the right people in NASA but when it came to chose the first of the astronauts to fly, much to Glenn's utter distress, NASA decided to do a straw poll among the very team themselves. When the team was assembled in NASA chief Bob Gilruth's office at the Langley Research centre, Gilruth named Shepard as number 1 on the launch pad. Shepard recalled that he stared at the floor, knowing that there were six very disappointed men in the room. He didn't want to look up, but when he did the six were smiling and good ol' Glenn was the first come up and shake his hand.

Shepard, Gus Grissom and Glenn were named as the three from whom one would be chosen to make the first flight, so the three had to keep secret the fact that after Shepard, Grissom would be up, then Glenn, who was to support both his colleagues' missions. These flights were simple 15-minute hops aboard a Redstone missile from Cape Canaveral.

As Shepard made his way to the pad on 2 May 1961 he remained anonymous to the public, his name had not been released. After a launch cancellation, his name was revealed. The next attempt was made on 5 May when Shepard was in the capsule for 4hrs 14min before it blasted off. During this time he was compelled to urinate but because highly optimistic managers assumed that Shepard would be launch on time and had not equipped him with a recepticle, the urine filled his spacesuit in the arch of his back. "I'm a real wetback now," said the astronaut, who was by this time getting rather impatient with what seemed to him to be over-cautious delays for this and that. Finally, he said firmly, "Why don't you fix your little problem, and light this candle". The launch proceeded shortly after, though, contrary to what the film *The Right Stuff* would have you believe, not as a direct result of Shepard's urging. Listening to Shepard as he flew the mission, you would think he was reading a prompt card. He had flown the mission so many times in a simulator that his scheduled reports were uttered in monotone, even an unscheduled utterance, "What a beautiful view!" sounded as though Shepard was trying to convince himself that it was, because all he saw was the Earth in black and white through his periscope. The flight of *Freedom 7* was perfect. After being fished out of the Atlantic after the planned splashdown, an ecstatically proud Shepard, said "Boy, what a ride". The hero was home. Tickertape welcomes, medals of honour from President Kennedy. It was great while it lasted.

John Glenn captured the imagination of the whole world with what was seen as the heroic flight of Friendship 7 into orbit and around the Earth three times. The high drama was even spiced with the "is his heatshield loose?" story. "Boy, that was a real fireball," said Glenn, his heatshield in fact not being loose. Newspapers started to describe him as America's first man in space. Shepard felt piqued. By the time Scott Carpenter, Wally Schirra and Gordo Cooper had also became orbital heroes, Shepard had been forgotton. This made him even more determined to follow Cooper's flight, which he backed up, with a 48hr effort in orbit. The flight was cancelled and, in 1963, Shepard went into training for the first manned Gemini flight, with either Frank Borman or Tom Stafford as his final choice as pilot.

It was at about this time that Shepard had started to suffer dizzy spells. Doctors found that he was suffering from Meniere's Syndrome, a build-up of fluid in the inner ear which results in vertigo and deafness. Shepard was grounded, not just from space but from aircraft, too. The bitterly disappointed astronaut must have fixed his eyes even more firmly on the Moon. He was determined to get there, to get the glory he deserved and the goal he had so naturally aspired to.

Shepard became the chief of the astronaut office, ruling it with a mixture of humour and discipline from his LSD – large steel desk. He was a power unto himself and could be tough on the lower ranks. It was Shepard who chose the crews for each mission, so it was rather important to stay on his side and to impress. Scientist astronaut Brian O'Leary, recalled that going to Shepard's office was like going to the headmaster for a ticking off. O'Leary said that in Big Al's presence he could hardly put two words together without stuttering.

Shepard perservered with his condition, keeping fit but turning his attention outside his NASA job to the Houston civic and business community, becoming the only astronaut to make his mark in this area. Investing his share of the $500,000 *Time Life* contract money that NASA had negotiated for *Time Life*'s exclusive coverage of the astronauts, he joined a banking partnership and within five years had become a millionaire, maybe six times over. The banks did not go out of their way to exploit Shepard's name but whenever he appeared at them during the early days, media attention was intense, gaining valuable publicity which earned one bank $20 million and made Shepard rich. Here was a man exploiting his astronaut status to the full and he hadn't been to the Moon yet. There were oil strikes and investments in a ranch. It was real Dallas stuff, this. For JR read ABS. Shepard drove a Corvette or Cadillac to work. He mingled with the rich and famous. He was, he said, "gathering nuts for the winter"[1].

But it was still an unhappy time for the frustrated Shepard. Shepard, along with his grounded sidekick, Deke Slayton, the Mecury man who never flew, continued to pick the crews – knowing he was better than the people he was chosing – always hoping that he could be one of them. By 1968, Shepard's condition was getting worse. "A man makes his own luck," said Shepard, who decided to have a risky operation, it was his last chance to make the Moon. Secretly under the pseudonym of Victor Poulis, in May 1968, he entered a Los Angleles hospital, for an operation, which he paid for himself, to have a small tube inserted into his inner ear so that the excess fluid in there would run off into his spinal column. The operation was a success. The already-fit Shepard entered into a period of frenzied activity in the gym and in simulators and in May 1969, when Apollo 10 was flying, he was restored to flight status.

By this time, Shepard was selecting the flight crews for Apollo 13 and 14. The commander of Apollo 13, the next available flight, was to be Alan B. Shepard. Many other astronauts, particularly the scientists, who had been training for a flight for years, were bitter and angry. Shepard could afford to ride over this, claiming that he had fought hard to get a chance to something that he thought was important, including going to the Moon. Pressure from some circles persuaded the 47 year old Shepard to move to Apollo 14 to give him a little more time to train. Thus James Lovell and his two crewmen were destined to become key participants in the Apollo 13 drama. As Shepard waited to blast off on Apollo 14, with a space crew with just five minute's space experience, he knew that he would either be the space hero he should have been from 1961 or, if the flight failed, the "son-of-a-bitch, who just wasn't ready". He was ready, the only Mercury astronaut to make the Moon. So was his 'golf club'.

Shepard in 1986. (from A. Shepard)

After his return, Shepard served as a delegate to the 26th United Nations general assembly and ruled the astronaut office from his LSD yet again until August 1974, when he left and retired from the US Navy with the rank of Rear Admiral. He felt it important to make a complete change and try to hang on the the good old days, he said. He became chairman of the Marathon Construction company which built K Mart shopping centres in the USA. In 1976, Shepard formed the Windward Company, a distributorship of Coors beer, and was president of this company until 1985. He is an investor in K Mart stores and Marriott Hotels, director of the Shaklee Corporation, American General Bond Fund and American General Convertible Securities. He is also on the board of directors of the Houston School for Deaf Children. Shepard's business isn't space any more, its business.

An avid golfer, Shepard plays many pro-Am tournaments around the world, cheerfully answering questions about his moonflight, particularly emphasizing the sight of the small Earth in the blackness of space. At those who ask a stupid question or those he doesn't like the look off, he fires his laser beam. He has never gone out of his way to seek popularity other than with a close circle of friends on his level. As described in Lawrence Wright's article 'Ten Years Later – the moonwalkers' many of those are his golfing pals, one of whom described how he was playing with Shepard when the moon man disappeared. His friend found Shepard grovelling in a waste bin, or trash can, muttering "six, seven, eight...." Shepard looked up with a grin, "Eight Coors, one Schlitz. That's not bad!" he said. Not bad for an old (and rich and happy) man.

Ed Mitchell – the Thinker

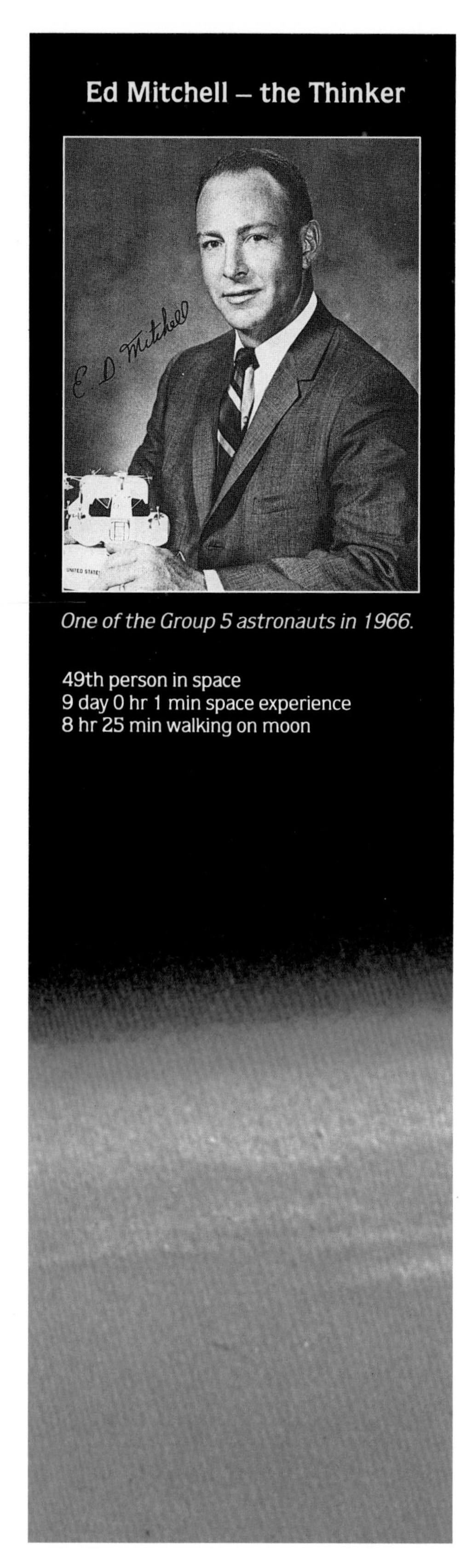

One of the Group 5 astronauts in 1966.

49th person in space
9 day 0 hr 1 min space experience
8 hr 25 min walking on moon

EDGAR DEAN MITCHELL IS PROBABLY THE most anonymous of the moonwalkers and he is quite happy about that. "I don't like the limelight," he says. But Mitchell is also unique amongst the moonwalkers. When he flew Apollo 14, he held a doctorate in aeronautics and astronautics and also held a degree in industrial management. Mitchell was also one of the best qualified fliers in the US Navy. Probably the most intellectual of the astronauts, he performed some experiments during the flight without the knowlege of NASA, who when it found out were none too pleased. Mitchell performed the first extrasensory perception experiments in space. Five symbols – a square, a cross, a star, a wavy line and a circle – were orientated randomly in columns of 25 and four of Mitchell's colleagues in the US attempted to guess the order of the symbols. No one else knew about the experiments until one of the four on Earth spilled the beans. The experiment, however, was deemed to be a reasonable success in ESP terms: 1 in 3,000. Reasonable success? Undetered by ESP and deterred by NASA, Mitchell left the space agency to establish The Institute of Noetic Sciences in Palo Alto.

Mitchell was born in Hereford, Texas on 17 September 1930 to Joseph T. Mitchell and the former Ollidean Arnold. The family moved to Artesia, New Mexico soon after and Mitchell went to primary school in Roswell and graduated from Artesia High School. He graduated from the Carnegie Institute of Technology in Pittsburgh, Pennsylvania with science degree in industrial management in 1952. During his period at Carnegie, Mitchell married the former Louise Randall of Pittsburgh on 21 December 1951. On 10 September 1952, he enlisted in the US Navy. Mitchell underwent recruit training in San Diego, California and then attended the Officers' Candidate School in Newport, Rhode Island. On 7 May 1953 he was commissioned as an ensign in the US Naval Reserve to begin flight training at Pensacola, Florida and at Hutchinson, Kansas. For achieving the highest overall marks in flight training, Mitchell received the Daughters of the American Revolution Award and was awarded his wings. On 12 August 1953 Mitchell's first daughter, Karlyn was born. Mitchell was then posted to the Fleet Airborne Electronics Training Unit of the US Pacific fleet from July to September 1954 before joining Patrol Squadron 29 deployed at Okinawa until November 1955. Mitchell was then transferred to Heavy Attack Squadron 2 based on the *USS Bon Homme Richard* and *USS Ticonderoga*. During this period, on 25 July 1956, Mitchell became a fully fledged US Navy Lieutenant after his commission in the Naval Reserve was terminated. In April 1958, he moved to Air Development Squadron 5 as project pilot of the A3D aircraft. His second daughter, Elizabeth, arrived on 24 March 1959.

Posing (right), before Apollo 14, with Shepard (centre) and Roosa.

From July 1959 to August 1961, Mitchell attended post graduate school at Monterey, California from which he received a degree in aeronautical engineering, attaining the rank of Lieutenant Commander, before moving to the Massachusetts Institute of Technology to graduate with a doctorate in aeronautics and astronautics in June 1964. Until April 1965, Mitchell was chief of the programme management division for the Manned Orbital Laboratory project. MOL was a Gemini-type spacecraft which would conduct military missions, flying a specialist group of astronauts not associated with NASA, but the project was cancelled in the summer of 1965 by which time, Mitchell had entered the Air Force Aerospace Research Pilots School at Edwards Air Force Base, California. He graduated first in his class and became an instructor at the school. In April 1966, having been promoted to Commander, Mitchell's name was among the 19 listed to become Group 5 NASA astronauts.

With several former astronauts in 1978. Left to right: Shepard, Schirra, Mitchell, Schweickart, Cernan, Gordon, Anders, Duke, Conrad, Aldrin, Lovell, Cooper, Evans, Scott, Armstrong, Worden, Irwin, Roosa, Stafford, Cunningham, Pogue and Collins.

A reflection of his impeccable qualifications was that, Mitchell, who served on the support group of Apollo 9, was the first of his group to be assigned to back-up a flight. He was back-up lunar module pilot of Apollo 10, serving with Gordon Cooper and Donn Eisele, naturally assumed to be what would be the Apollo 13 crew, as a result of the usual NASA-style crew rotation system. Things didn't quite go that way this time, however. Shepard made it back to flight status, bumped-off Cooper and Eisele and chose Mitchell and Stuart Roosa to be his crew on

Apollo 13, which became Apollo 14 when he was persuaded that he needed a little more time for training.

If the truth be known, Mitchell was a reluctant astronaut and was getting fed up with the shenanigans and with NASA bureacracy. Prelaunch restrictions were a 'blasted nuisance'. He told astronaut chief Deke Slayton that he would leave NASA after Apollo 14. Slayton told him that after Apollo 14 he would have to back-up Apollo 16, a dead-end assignment since Apollo 19 had already been cancelled, and if he refused, he wouldn't fly Apollo 14. Mitchell flew Apollo 14 and backed-up Apollo 16! He was rather a portly astronaut and the subject of some jokes, one of which was him being handed a pair of size 52 overalls before he entered Apollo 14 with a message: "Ed, please watch your diet". Mitchell was overshadowed by Alan Shepard and was none too pleased when he was proved correct in his assertion that he and Shepard were much nearer to the rim of Cone crater than they thought, and should have pressed on. Mitchell's interest in psychic phenomena led him to perform the ESP experiments during rest periods of the the flight and his Moon experience is indelibly

linked to the area of parapsychology.

He told the *Observer Magazine*: "Looking out the window of the spacecraft, at this little planet floating out there, I began to have some rather startling insights and feelings. It was as though all of my previous philosophical questioning about the purposeful universe, as opposed to a purposeless universe, was suddenly resolved. I had a deep knowing that this was not just a random matter floating out there, that is was party of a divine plan – not divine with a capital D, but ordered, structured, purposeful." He believes that the most important part of Apollo was the act of going out and returning safely, from seeing the planet as it really is. It has caused mankind to see itself differently than before. He told *Omni* magazine that he had a 'religious' experience, a realisation that the Universe is spirit and matter but they are not separate. The bridge is consiousness. God is something like a universal consiousness manifested in each individual.

In *Spaceline*, a publication of the Institute for Security and Co-operation in Outer Space, of which Mitchell is chairman of the board of advisors, Mitchell wrote: "As we manage the physical, the technology, the science, it is simply the expression of our inner awareness. Thus, as we become more aware, we will create a different technology. But clearly, from the subatomic to the cosmic level, there is a harmony, there is a pattern, there is a repetition of the same process over and over on a grander scale. So, the subatomic, the microscopic world, and the cosmic, the macroscopic world, are all interconnected. It must be viewed much closer to that of a living, thinking, intelligent organism on the cosmic scale. Scientific knowledge and religious knowledge have never been satisfactorily integrated. They're just two separate bodies of knowledge each of which seems to be able to hold part of the truth, and they think they hold all the truth, but we've never been able to get them together. The insight that occurred to me on Apollo 14 as I looked at planet Earth was indeed that it was now time to bring these two bodies of thinking together. That, indeed, the religious traditions we have been living with for two thousand years or more, both in the East and West, needed some adjustment. They weren't really expressing reality as it exists. That it was now time to evolve those bodies of divine and spiritual thought into a form compatible with science, and to change science in a way that made it compatible with divine ideas."

Life's not so bad after the Moon.

After Apollo 14, the now-divorced Mitchell had a frustrating period having to fill in time as Apollo 16 back up LMP, with commander Fred Haise and CMP Stuart Roosa. Haise had replaced Shepard who presumably felt that he was above such things. Mitchell was itching to get into the world of ESP and like things. In early 1972, there was a chance that he would getting into the world of the Moon again, for the prime LMP Charlie Duke wound up in hospital with pneumonia weeks before the flight. Duke recovered and flew, mainly because Apollo 16 had been delayed. On 1 October 1972, Mitchell was let out of NASA and the US Navy. He immediately set up E.D. Mitchell Associates, in Palo Alto, California, to write a book, and the Institute of Noetic Sciences to study the 'nature of consciousness', working with researchers at Stanford Research Institute.

Mitchell's interest in psychic research began as a search for the meaning of life, concepts that he had not found in religion or philosophy. Psychic research appeared to offer a rational and substantial support for many theological and philosophical concepts and explained why people throughout history have been claiming a spiritual foundation to the physical world. As he wrote in the *New York Times* "Evidence from my research suggest that awareness can operate beyond the body and therefore it is not unreasonable to hypothesize that mind may be able to operate independently of the body. Evidence of mediums, as documented by eminent scientists performing survival research indicates this possibility. Death may simply be an alteration in consciousness, a transition for continued life in a non-material form." Such thoughts from the sixth man on the Moon are enough to make some scientists and men of religion freak out and Mitchell is aware of this. But as new science is developed, we must think differently, he says, to a degree that perhaps we may understand these things instead of relegating the lot to mystical god faces just because we are incapable of understanding. If mind can influence matter, then mind is not the product of matter.

Being a moonman is a good marketing tool and Mitchell used this to apply his research to human resources management, moving to West Palm Beach, Florida to set up E. D. Mitchell Corporation, EMCO, in 1974. This involved selling programmes to teach people to make decisions intuitively, not emotionally, i.e., to teach them to trust their hunches. This led him, in 1978, to establish, with two Mormon partners, Forecast Systems Inc. Mitchell has remarried and adopted three more children from his wife's previous marriage, two girls, Kimberley and Marybeth and an only son, Paul. "I currently live very quietly, without a secretary or PR firm. He is writing three books to leave to posterity "my impressions of the who, what and where aspects of evolution".

And of ESP he now says, "I don't like it, it's just hype".

Dave Scott – the Entrepreneur

The first Class 3 astronaut to fly; Scott in 1963.

25th person in space
10th person to make two spaceflights
8th person to make three spaceflights
22 day 18 hr 53 min space experience
17 hr 36 min walking/driving on moon

DAVE SCOTT'S APOLLO 15 MISSION WAS the pinnacle of the American achievement. It was the "most complex and carefully planned scientific expedition in the history of exploration", said NASA. After Apollo 16 and 17, which were essentially duplicates of Apollo 15, nothing of comparable challenge, risk or reward has emerged since. Apollo 15 could be seen as the maximum a human team can ever successfully achieve. Yet, what is Dave Scott remembered for? The Envelopes! The envelopes saga was for the most part, a media myth, concentrating on a non-issue of Apollo 15, just as it seemed, to the astronauts, that the media concentrated on the 'non-issues' of the whole programme. "The media coverage of Apollo was crap!" said one of the moonwalkers.

The story of Scott's envelope saga goes, essentially, that he smuggled contraband envelopes among those that had been authorised aboard Apollo 15, franked them on the Moon, gave some of them to a business acquaintance, Horst Eiermann, who sold them, giving Scott and his crew, Irwin and Worden, a little nest-egg for their children. When NASA found out, the story goes, the crew was fired by astronaut chief Deke

Straight in to Gemini 8 in 1966, with Armstrong, left.

Slayton. In fact, many 'non-mission' items were carried on every US manned flight. Not just the astonauts but a 'team' was involved. If there was any problem in the past, the 'team' disappeared, leaving the crew very much exposed. It happened before Apollo 15: for example, the Apollo 14 crew took medallions aboard, and certain post-flight activities involving these caused internal ruc-

After Apollo 9 in 1969. Scott (centre), with Schweikart (left) and McDivitt.

After Apollo 15. Scott (right), with Worden (left) and Irwin.

Scott of Hadley Base.

tions which never surface in the media. The luck ran out with Apollo 15.

Each of the envelopes carried on Apollo 15 was prepared, franked, packaged and stored aboard the spacecraft with the participation of numerous NASA personnel. Horst Eiermann was in fact introduced to the Apollo 15 crew by astronaut chief Deke Slayton himself and, after Eiermann's involvement with some of the envelopes, the crew did not receive any money from him. But the 'team' had got careless with some of the envelopes and it became clear, some months after Apollo 15, that as a result of misinformed leaks about trouble in the astronaut corps, NASA had badly mishandled the envelope situation. Indeed, it admitted that it was "not proud of its performance". There had been no impropriety by the crew, in context of the fact that what they did was what others had done before. After Apollo 15, Scott, Worden and Irwin were assigned to a dead-end job backing up Apollo 17 and decided to move on to more demanding and interesting employment, though Scott stayed with NASA.

Scott hails from Texas, where he was born in San Antonio on 6 June 1932, the son of Tom, who became a Brigadier General in the US Air Force. He was educated at Western High School in Washington DC and later attended Military Academy at West Point, graduating the fifth in his class of 633, with a science degree. Not surprisingly, given the filial connection, Scott entered the US Air Force, completing flight training at Webb Air Force Base, Texas in 1955. Scott then reported for a gunnery training course at Laughlin Air Force Base, Texas and Luke Air Force Base, Arizona. After this, in April 1956, Dave Scott was posted to the Netherlands and the 32nd Tactical Fighter Squadron.

In June 1960, Scott entered the Massachusetts Institute of Technology to gain a masters degree in astronautics and aeronautics, with a thesis on interplanetary navigation. Two years later, just after Scott Carpenter had completed the second Mercury orbital flight, Scott was on his way too, via the Air Force experimental Test Pilot School, his eyes fixed firmly on NASA. Married to the former Ann Lurton Ott of San Antonio, daughter of another Brigadier General in the USAF, Scott began training at Edwards Air Force Base.

Becoming a test pilot a year later, Scott then entered the Aerospace Research Pilot's School at Edwards and almost met his death.

Scott in the late 1970s.

Flying in the front of an F-104B jet, simulating landings of the X-15 rocket plane, Scott faced a life or death decision. The aircraft lost control at the last second of landing and careered towards the runway. Behind him, fellow pilot Michael J.Adams, also an ARPS student, elected to eject. Scott stayed with the plane, making a crash landing. Had each made the opposite decision they would have been killed. Ironically, Michael Adams went on to fly the X-15 rocket plane, reached the edges of space in 1967 and was killed when the rocket plane went out of control and plummeted to Earth, breaking up on the way down. In October 1963, Scott had two pieces of good news: his son Douglas was born and he was requested to report for duty at Houston as one of the 14 astronauts in the third class at NASA.

Scott was assigned to specialize in spacecraft guidance and control systems and by June 1965 was acting as capsule communicator for Gemini 4. The same year, he became the first in his astronaut class to be assigned a mission, on the prime crew of Gemini 8, without having to serve as a back-up. Gemini 8 was to be a whizz of a mission, a rendezvous and docking, which, after Gemini 6's target rocket exploded, became the first attempt, with a spacewalk around the world, by Scott himself. The assured Scott and quietly assured command pilot Neil Armstrong, ex X-15 pilot, took to the skies after a 48-hour delay, on 16 March 1966. Rendezvous was effortless and, without even waiting to come into contact with a tracking ship, the *Rose Knot Victor*, Armstrong went ahead with the docking. Soon after, but for their intense training, he and Scott could have been killed. Unbeknown to the quietly jubilant crew a thruster on their spacecraft was about to fire intermittently due to a short circuit. They began passing signals to the Agena target, when suddenly the combination went into a spin. Thinking that the fault lay in the Agena which was perhaps misinterpreting the signals being sent to it, Armstrong elected to undock. The spinning got worse, reaching 70 rpm, and there was the added danger that Gemini would strike the Agena and explode. Less well-trained astronauts would have blacked out by now and the mission would probably have been doomed. But the crew shut down systems, including the stuck thruster and fired the re-entry control system thrusters, meaning an immediate abandonment of the flight and no spacewalk for Scott, who was to have become the second American to make an EVA. Landing was achieved safely after just ten hours in space but the recovery was uncomfortable, the crew having to wait in heaving seas for many hours, getting quite horribly seasick. But they had the compensation of being garlanded by Aloha girls in Hawaii, *en route* for the debriefing at Houston.

Clearly, Scott was impressing, for no sooner had he landed that he was being named as back-up senior pilot for the first manned Apollo mission. Shortly after this, the McDivitt, Scott, Schweickart crew was posted to a prime crew of Apollo 2B, to attempt the first flight of an Apollo lunar module. This then became Apollo AS-504, which could have become Apollo 8 taking the crew around the Moon, but at the insistence of commander McDivitt, became Apollo 9, riding the giant Saturn 5 just three years after Gemini 8. The Apollo 9 mission made an enormous contribution to the success of Apollo 11 but was rather overshadowed since it was taking place in Earth orbit and not in lunar orbit, as Apollo 8 had done. During the flight, Scott, flying solo in the command module *Gumdrop*, was the first to do so since Gordon Cooper in 1963. The classic, connoisseur's mission, also featured a stand-up EVA by Scott leaning out of the open hatch of *Gumdrop* to retrieve samples, while being photographed by Schweickart completely outside the lunar module. His photo of Scott remains a classic of the Apollo programme.

The ten-day mission completed, Scott then began his first stint as a back-up pilot, six years into a spaceflight career. As back-up commander of Apollo 12 he seemed assured of command of Apollo 15 and so it was, despite some rumblings that pressure to fly the geologist Jack Schmitt on Apollo 15 could result in the entire crew being replaced with what was to have been Apollo 18's crew, which was, ironically, backing up Scott, Worden and Irwin.

The Apollo 15 flight was a patriotic and decidedly Air Force affair. The ouspoken Scott, when asked what he wanted on his flight emblem, remarked, "what's wrong with red, white and blue". The mission was being flown by a Major, Lt Colonel and Colonel – Scott – in the US Air Force and began in the sultry skies over the Kennedy Space Centre in July 1971. There were many times during the flight that Scott was decidedly ratty. On the Moon he was the urbane commander, demonstrating the Galileo experiment, with a falcon's feather and a hammer being dropped together and landing at the same time. "Hey, Galileo was right!" chortled the colonel.

Apollo 15, which featured the first drives by a lunar roving vehicle and spectacular scenery of mountains and valleys was, for many observers, the pinnacle of the Apollo programme. It ended with a slightly bigger splash than usual, beneath two parachutes instead of three. The most exciting moment of the mission for Scott was when "Jim and I had driven up the slope of Hadley Delta. We were suprised when we turned round and realised how high we were. We were at a point we could never have reached on foot. I saw the LM more than three miles away. Taking it all in, our LM, the rolling plains and the mountains, I began to feel at home in our new surroundings". Scott said that when he had landed and looked back at the Moon his immediate reaction was to feel "a little homesick – you really get attached to it, you really do," he said[1]. Scott then began training to support Gene Cernan on Apollo 17. Then came the 'envelope sage'. Scott stayed with NASA which employed him gainfully for another six years.

First, Scott, who was on friendly terms with many cosmonauts, helped to support the Apollo Soyuz missions, making the first forays into the Soviet Union to prepare the ground. After the flight in July 1975, Scott became deputy director of NASA's part of Edwards Air Force Base, the Dryden Research Centre, in California. He resigned from the US Air Force in March 1975 on being named as director of the centre, an extremely prestigious appointment to match an eminent astronaut career so sadly spoiled by an indiscretion. Scott played a major role in getting the Space Shuttle rolling, overseeing Dryden's responsibility during the first atmospheric glide tests, called ALT, in 1977. Five flights of the Shuttle orbiter *Enterprise*, released from the back of a 747 aircraft, later, Scott, his work done, left NASA to enter private business. This, initially, was as Scott-Preyss Associates before becoming president of Scott Science and Technology in Los Angles and Lancaster, California. He did stints as the TV studio expert for Shuttle flights, performing admirably on ITN's coverage of the first launch in 1981, outdoing the BBC's laid-back reaction hands down.

The aim of Scott's company is to apply space technology to other areas of industry beneficially, which illustrates what the Moon flight did for him, something which he recalled in a *New Scientist* publication in 1986: "Clutching the ladder, I raise my eyes from the now familiar moonscape to Earth, glowing in the black heavens. That incredibly vivid sphere is so blue, so beautiful, so beloved. It is also bedevilled – by a threatened ecology, by starvation, by a shortage of energy that may motivate us to seek resources beyond the Earth. Our Apollo crew believe that technology capable of exploring space can and will help to resolve these problems. We feel a sense of pride in the accomplishments of our programme. Yet, we cannot escape a sense of deep concern for the fate of the planet and species."[2]

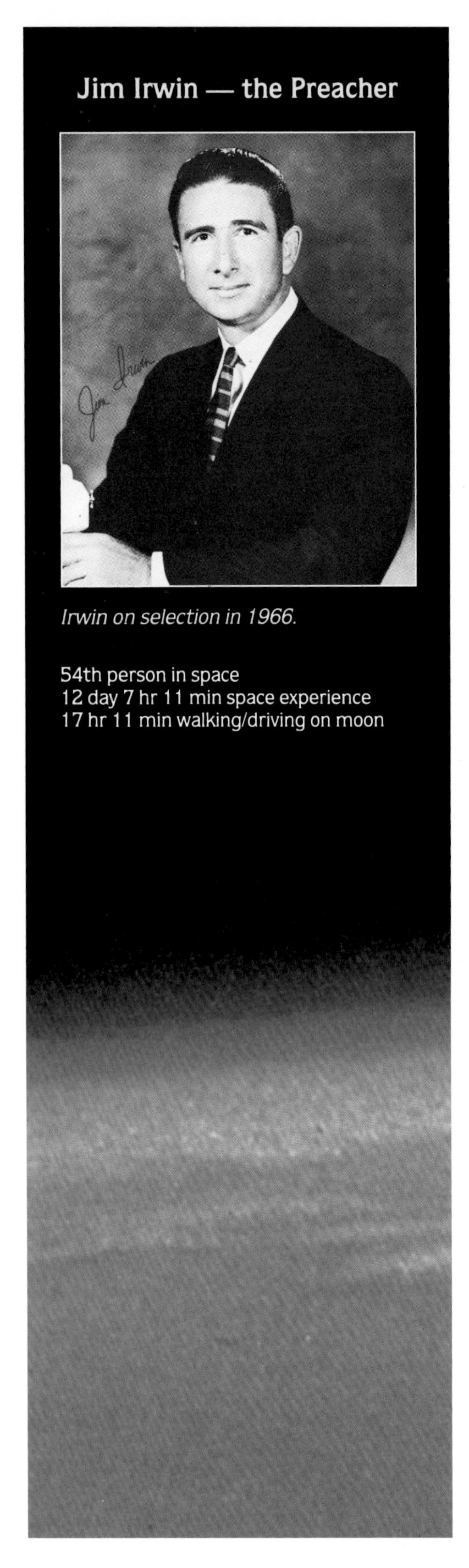

Jim Irwin — the Preacher

Irwin on selection in 1966.

54th person in space
12 day 7 hr 11 min space experience
17 hr 11 min walking/driving on moon

THROUGHOUT HIS AIR FORCE CAREER, James Benson Irwin wished he had left and become an airline pilot, retiring for the quiet life at 65. But "the Lord wanted me to go to the Moon so I could come back and do something more important with my life than fly airplanes." During the flight of Apollo 15 he "felt the beginning of some sort of deep change taking place within me"[1]. He had sort of drifted into the Air Force, wasn't the best of pilots and was smashed to pieces in an air crash in 1961. He also survived a series of heart attacks after he left NASA. The crash didn't stop him applying for the astronaut corps and the cardiac arrests haven't stopped his zealous quest to preach the gospel of the Lord all over the world. Lying in a hospital bed after his umpteenth heart attack in 1986, he was the only moonwalker who taped replies to my written questions in preparing this book. Some sent back blank tapes and presumably my request to the others never arrived.

Irwin considers his hometown as Colorado Springs but he took his time getting there. He was born in Pittsburg, Pennsylvania on 17 March 1930 and grew up in New Port, Richey and Orlando, all in Florida, Roseburg in Oregon and finally Salt Lake City in Utah. Irwin described his early years in his autobiography, *To Rule the Night*. He became a committed Christian at the age of 11 when still in Florida but, by his own admission, had become very much an uncommitted one when he reached adulthood. Occasionally he would drift back, then drift away again, he said. Irwin graduated from East High School in Salt lake City and attended US Naval Academy in Annapolis, graduating with a science degree in 1951. He certainly didn't want to become a sailor, and had almost drifted out of the academy. By chance his name came up in the draw of 20 per cent of the students who could be taken by the US Air Force. He received flight training at Hondo Air Base and Reese Air Base, both in Texas. He was posted to Yuma Air Base, Arizona as a fighter pilot, starting on piston-engine aircraft then graduating to T33s and then the B45 bomber. He married Mary Etta Wehling, a devout Roman Catholic, who divorced him shortly after, the result of religious conflicts more than anything else. In 1955, Irwin had his heart set on becoming a test pilot – at, of course, Edwards Air Force Base. He had to earn some degrees first, so went to the University of Michigan, joining class mates Edward White and Jim McDivitt.

In 1957, the year of Sputnik, Irwin graduated with degrees in aeronautical and astronautical engineering, hoping to be Edwards-bound. But he was posted to Wright Patterson Air Force Base in Dayton, Ohio to be project officer for the GAR-9 nuclear missile to be launched from an F-108. Irwin was dropped from instructor pilot status for disciplinary reasons, after pretending that he and not his brother had been at the controls of a T33 when it crashed after an illicit assignation. To gain the status again, Irwin had to log in a number of student pilots and it was with one of these that he came near to being killed three years later.

During a visit to his parents in San Jose, in 1958, he met Mary Ellen Monroe of Corvallis, Oregon, a receptionship at a photographer's studio. Monroe's family were strong Seventh Day Adventists and opposed the ensuing relationship on the grounds of religious incompatability. "Why was I always plagued with religious problems?" asked Irwin. Love conquers all and he and Mary finally married, having their first child Joy in 1959, the year the Mercury Seven were chosen. In 1960 Irwin finally made it to Edwards Air Force Base and the experimental test pilots school, where Tom Stafford was an instructor and Frank Borman and Mike Collins fellow students. He graduated in 1961, the year that his second daughter, Jill, was born. It was about this time that he realized that he could fly into space, perhaps to the Moon, as a result of the flights of Shepard and Gagarin. As a child he once had a dream in which he walked on the Moon but didn't remember that he had until reminded by his mother when he became an astronaut. Irwin was assigned a secret posting as director of the test force of the ASG-18 and GAR-9 weapons systems.

During a flight at a local Aero Club, working to become an instructor pilot again the student fouled up and the plane crashed in the desert. "I hit the front seat with the side of my head, I was wearing tennis shoes and the front seat collapsed on my feet, particularly the right one. It came down above my ankle, giving me a compound fracture, with bones sticking out through the flesh. I had two broken legs, a broken jaw and head injuries which gave me amnesia. They nearly amputated my right foot"[2]. At the time he had just been told that he was to be the first and only pilot of the YF-12A, the

Chosen for Apollo 15, right.

fastest and highest thing around with wings. It took Irwin five months to recover and then he was given the crushing news that he had been grounded for a year because of his head injury and experience of amnesia. While Frank Borman and Edward White were graduating from the Aerospace Research Pilots School, Irwin was left fretting.

At last, in late 1962, Irwin was flying again and entered the Aerospace Research Pilot School at Edwards. During one F-104 horror flight, Irwin suffered control loss at 90,000 feet and managed to control the plane with only 3,000 feet to go. He graduated from the school and applied to join the third group of astronauts in 1963. He was turned down but at least that year he had a son, James. Irwin got to test the YF-12A and in 1964 had another daughter Jan. He was turned down again by NASA when he applied to join the fourth group, who were in fact 'scientist' astronauts. He was, however, assigned as chief of advanced requirements branch of US Air Force command at Colorado Springs. In 1966, with one month before being too old to apply to become a NASA astronaut, Irwin applied and was chosen with the 19 fifth group. "I never thought that so many Moon missions were being considered, I reckoned on perhaps two or three."[3] By specializing in lunar module systems, Irwin reckoned he had a better chance of making it to the Moon. Irwin was crew commander of the LTA-8 lunar module thermal vacuum test in June 1968. After this he was approached by Tom Stafford who was to be named as commander of Apollo 10. Thinking he would be named back-up lunar module pilot, he ended up on the support crew. Fellow fifth-class astronaut Ed Mitchell was to be back-up. "This was a blow to my ego," said Irwin. The irony, Irwin noted, was that Stafford was a year behind him in Naval academy! During support of Apollo 10, Irwin was approached by Dave Scott and asked whether he would like to join him on the back-up crew of Apollo 12.

As Apollo 15 loomed Irwin's marriage was on the rocks. The emotional turmoil at home almost caused Irwin to go to astronaut chief Deke Slayton and request removal from the crew, he says. Mary and Jim even discussed divorce. It wasn't the best of ways to start a mission in which you were to become the eighth man on the Moon.

Irwin found God on the Moon. "I had become a sceptic about getting guidance from God, and I know that I had lost the feeling of His nearness. On the Moon the total picture of the power of God and His son Jesus Christ became abundantly clear to me. I felt an overwhelming sense of the presence of God on the Moon. I felt his spirit more closely that I have ever felt it on the Earth, right there beside me. It was amazing, I didn't change my habits. I prayed at the same times that I do on Earth, a brief prayer before I go to sleep and then when I wake up. But through those days there was gradually enhanced feeling of God's nearness. And when we were struggling with difficult tasks on the first EVA, when a key string broke and I couldn't get the station up, I prayed. Immediately, I had the answer. It was almost like a revelation. God was telling me what to do. I am not talking about some strange sense of direction. There was this supernatural sensation of his presence[3]."

Irwin explained that he was not the only astronaut to have this experience, all had to some degree been effected by it. "Everybody felt that they were more efficient, that they had achieved a feeling of mental power. We all thought with a new clarity, almost a clairvoyance. I could almost anticipate what Dave Scott was going to say, I felt I knew what he was thinking." Irwin was so inspired by his experience that he quoted the beginning of Psalm 121 despite the fact that he didn't know it word-for-word: "I will lift up mine eyes unto the hills, from whence cometh my help". This was the only indication people had on Earth that Irwin was even thinking about God, let alone had found him.

Back on Earth, Irwin felt a new man. He described these new feeelings to me in taped answers to written questions. He felt his life was beginning again but couldn't understand why. He knew that things were going to be different, for a start the crew would belong to NASA for the next weeks of hectic tours and speeches. When Irwin thinks of Apollo 15 he immediately places himself back in the valley at Hadley, hemmed in on three sides by those 15,000 feet mountains. "It seemed a place that I was meant to go," he said. "I always had a love for mountains and enjoyed climbing them and was glad I got to climb these too, at least a bit". Irwin also thinks of the Earth from space, "a compelling, beautiful object in space, no matter where you view it from and in whatever lighting conditions". He saw it as the size of a basketball, the size of a baseball, then a marble. "It was the full Earth to start with, then the half Earth while we were on the moon, then it became a crescent, then was just in darkness, in moonshine. We had beautifiul changing views and that was a great opportunity and a great blessing." He said later, "we saw the earth as God sees it and realized what a special place He has made for us. The flight gives me a lot of perspective, a new perspective of the Earth and a new responsibility towards it". Irwin also said openly at the post-flight press conference that he felt the presence of God on the Moon "in a very profound way". This religious experience was the most important thing that had happened to him but he didn't know what to do with it. As he travelled, making speeches and giving talks to all sorts of people of all ages, Irwin gradually became aware of a religious or spirtual awakening. Eventually, communicating his religious experience on the Moon graduated to him addressing 50,000 people in the Astrodome, something that before the flight would have completely terrified the quiet man. There was an "explosion of interest," he said. "I have never gone out of my way to create the celebrity status, although the public would never want to regard the twelve moonwalker as anything less. The point I make to people is that because of the gift of Christ we are all celebrities in Gods eyes. The difficulty I had in adjusting to this was speaking to people in public and sharing my thoughts with them. I was very shy, much happier when I was at the controls of a plane or spacecraft, but I was being taken from talking about technology to talking about human feelings and finding the right words to use. I have a limited time with these poeple and I want to say something to them that convinces them to change their view of Christ and accept him.

With the rover at Hadley.

I accept that notoriety is inevitable in my case and that I need it to do the work I do for the Lord." He got busier and busier, preaching the faith through his experience. He decided that this was the direction in which he must go and announced that he would retire in July 1972. There were, however, a couple of little problems. First, he was on the back-up crew for Apollo 17; a dead-end job, because even if Schmitt broke his legs, NASA would wait to fly him. Secondly, there was the matter of the unauthorised envelopes that

The evangelist. (Carlson Photography)

were franked on the Moon during the mission.

By this time Irwin had decided to give NASA a miss anyway. "Because I had already retired I was in a position to be able to say more about the issue and I admitted that we had made a mistake and that we were only human. Dave and Al were not quite as open and there was a bit of bad feeling. I wish we were as close in our crew as we once were, may be a reunion for just us three together. But it hasn't happened," he said later.

Irwin started his new life, establishing the High Flight organisation and travelling the world, preaching the gospel through his experiences in Apollo 15. He went to Saigon, Vietnam, where his wife, Mary, fell in love with a little orphan whom the Irwins adopted as their fifth child, Joe. On they went, to Vietnam, Taiwan, Australia, New Zealand and then to the Holy Land. During his tours he said, "Before the flight I was really not a religious man, I believed in God but I really had nothing to share. But when I came back from the Moon I felt so strongly that I had something that I wanted to share with others. I established High Flight in order to tell all men everywhere that God is alive, not only on the Earth but on the Moon". Apollo 15 changed Irwin physically, psychologically and spiritually. He came back with an appreciation of things that he had previously taken for granted and an appreciation of other people and with a new regard for God and a desire to serve Him. Irwin says that it was "a great opportunity to fly Apollo 15 and I was sent there by the people of the Earth and I have a responsibility to those people to enrich their lives the way mine has been enriched as a result of the mission. I have become a servant to lift them up, not to the Moon but, in a vicarious way, to lift them in a physical and spiritual way." Irwin's schedule was more hectic than the tour made by the three Apollo 11 astronauts after their flight. The schedule for 1973 and 1974 was getting pretty full. Something had to give and it turned out to be Irwin's heart.

Playing a mean game of tennis in Denver on 4 April 1973 he had a heart attack. "I was grounded again," said Irwin afterwards. It is quite possible that Irwin's heart attack began on the Moon. He had a history of slightly high blood pressure anyway, and his exertions on the Moon were reflected in extremely high heart rates. This was one reason why the Apollo 16 crew had orange juice laced with potassium. In 1977, he had bypass surgery, only to suffer a second heart attack later that year. Irwin's High Flight quest continued with five trips to Mt Ararat to find the remains of Noah's ark. During one of these on 20 August 1982 that he fell from a cliff and was in hospital for several days.

Two weeks after announcing that he planned another trip, on 7 June 1986, Irwin was out jogging in Colorado Springs. He felt exhausted, sat down on a pavement and suffered another heart attack. He was found by a boy who phoned an amulance. Irwin's heart stopped and needed a defibrillator to shock it back into some sort of action. He was in critical condition for several days but left hospital on 22 June to continue his work for the Lord.

Apollo's significance to Irwin seems greater each year but more in terms of how America managed to harness and direct that energy into the programme and of how successfull it was, particularly in light of the *Challenger* disaster. "We were really blessed during Apollo. I was greatly saddened by *Challenger* and it seemed that, when there were the other failures in 1986, God was somehow closing our door to space in a very dramatic way, causing us to realize that perhaps we had become over confident and to realize that we cannot go on expecting the

Suited-up and raring to preach. (from J. Irwin)

successes and treating it as commonplace and routine and that we can do something anytime we want to. I hope that what results will be another Apollo but I don't think any programme will ever capture the attention and imagination of the world, not even a Mars flight. We must pray for success when we travel into the arena of space."[9]

Irwin feels that although the other eleven meet on very rare occasions and are not great buddies but just friends, that each felt that their lives were changed by wallking on the Moon. "I think there has been an inner change. They appreciate the Earth more and have experienced a spiritual change but some may care not to admit it and accept the fact."[10] Five of the moonwalkers met on top of a mountian in Tennessee for a spirtual retreat with Irwin. He asked all to attend but only got four takers. "I found it significant and they did too". I believe we are obligated to the human race to tell everyone what happened to us and that our lives were richly rewarded through the spiritual relationships, maybe as a result of people praying for us during our missions. We felt the power of prayer, we were the focal point of prayer power. We are all in a position to travel the world and tell them what it was like and to thank them for the experience." Tell that to Armstrong.

John Young – the Astronaut

On selection in 1962.

18th person in space
5th person to make two spaceflights
3rd person to make three spaceflights
2nd person to make four spaceflights
1st person to make five and six spaceflights
34 day 19 hr 42 min space experience
20 hr 14 min walking/driving on moon

IF THERE IS ONE OF THE TWELVE WHO will keep his spaceboots on until the day he goes to THE big space station in the sky, it is John Young. He'd probably like his tombstone to be engraved with the words, "veteran of twenty spaceflights" but it looks as though he'll have to settle for six. Even so, that is still more than anyone else, so far at least. This number of flights could not have been predicted earlier in his space career when he was the first astronaut to be publicly reprimanded and one of the few to divorce and remarry. He did so quietly, before Apollo 16, that many thought when he landed that he was still married to his first wife.

Young lives, works, sleeps, eats and breathes spaceflight. He always has; since joining NASA in 1962. He has few interests outside space and typically his only exercise is sometimes listed as working-out in a full pressure suit. Young is a loner, a man of few words and those are usually awkwardly put together and delivered in a mumble. His smile is quite extraordinary. When you talk to him, every now and again he flashes a one-second long smile which oozes with shyness. Young has a gangly gait and a dishevelled look about his face, like someone who's just woken up. He doesn't discuss his feelings, especially about the Moon experience. "I think the moon's a pretty interesting place," is about all you'll get from the sixth man on the Moon. But he does have a delightful deadpan sense of humour which is easily lost in his drawl. Describing on TV the likelihood of a hydrogen fire spreading up the side of the Space Shuttle after a launch pad abort, Young said that a crew abandoning the vehicle across the

With Mike Collins before Gemini 10.

Preparing for Apollo 10.

Commander Apollo 16.

Leaping salute at Descartes.

walkway on the pad, "would get a bit crispy on the way out". Careering down the runway at Edwards Air Force base at the end of the first Shuttle mission, Young said of the unpowered Orbiter *Columbia*, "do you want me to put it in the hangar?" Aboard *Charlie Brown*, his Apollo 10 command module, he floated upside down, his head being pushed up and down by his commander Stafford. "I just do whatever he says," said Young. When a NASA geologist suggested that Young scrawl some messages on the lunar soil during Apollo 16, he replied, "I'll stomp out any words you want, except 'help!'"

Although Young is still an active astronaut, at the age of 58, he has been pushed to one side. No longer chief of the astronaut corps, he has become a technical assistant. He had been named to command the Shuttle mission to deploy the Hubble Space Telescope but when the crew was reconfirmed in 1988, after NASA had recovered from the *Challenger* disaster, Young was not on the list. He had become a casualty of the *Challenger* disaster, for speaking out loudly and angrily after the explosion about how safety had been compromised and further crews could be killed if safety wasn't put first. After he had emerged from the Shuttle *Columbia* after the first mission in 1981, he jumped around the undercarriage, punching the air with absolute delight and pride. "The way to the stars is open," he said optimistically. The space bubble burst for Young on 28 January 1986 when *Challenger* became the first manned spacecraft to take-off and not reach its destination. *Challenger* and its aftermath appeared to have a dramatic effect on Young's mood and personality.

John Watts Young was born in St Luke's Hospital, San Francisco on 24 September 1930. His father was a sales rep for a building company and was forced by the depression to send his family to his home state of Georgia. Young went to Cherokee Avenue School, Catersville starting in 1935, where he was described as bright and brilliant. His early years were described by Dave Shayler in *Spaceflight* magazine. After living with his aunt for several years, Young and his younger brother Hugh were moved to Orlando in Florida where their father had found a job in 1936. However, before Mr Young could find a suitable permanent home for them John and Hugh Young had to travel back and forth between Catersville and Orlando, finally settling in Orlando in 1939. Young became absorbed with aircraft, watching many types flying from nearby air bases. He started to build model aircraft, did a paper round and was a quiet but good pupil at school. His father served in the Second World War with the SeeBee's in the Pacific. In his teens, Young developed into a brilliant sportsman, keen on handball and running as well as football. Despite his quiet nature, he was regarded as a born leader. At the age of 16, Young, already a Buck Rogers fan, developed an interest in rockets and even delivered a paper on the subject at school. He graduated from Orlando High School in 1948 and went to the Georgia Institute of Technology under a Navy Reserve Officer Training Course. Four years later, Young earned his degree in aeronatical engineering, became a Navy serviceman and applied to fly. Before he could do this, however, he was assigned a year's tour of duty on the *USS Laws* as a fire control officer. His days as a 'black shoe' were the subject of some mirth in the astronaut corps. Young entered flight training in June 1953 at the Naval Basic Training Command at Pensacola, Florida. A year later, Young entered the Navy Advanced Training School at Corpus Christi, Texas and then moved back to Florida and Jacksonville Naval Air Station. In 1955, he was assigned a four year tour of duty flying with Fighter Squadron 103 from Jacksonville and from the *USS Coral Sea* and *USS Forrestal*. Also in 1955, Young married Barbara Vincent White of Jacksonville. His daughter Sandy was born in 1957 and a son, Johnny was born two years later.

Suiting up for flight 5, the first Shuttle mission.

In February 1959 Young made the logical move: to the US Naval Flight Test Centre at Patuxent River, passing his test pilot course and flying operationally from the base, testing various weapons systems. In 1962, as America was going wild about the flight of John Glenn, the enthralled Young began working on the High Jump project. This involved flying a Phantom F4B, parked on the runway at Naval Air Station at Brunswick in Maine, with engines running, to a height of 3,000 metres (9842 ft) in just over 30 seconds. A second flight in April aboard a Phantom F4H1, from the Point Magu Naval Air Station resulted in Young reaching 25,000 metres (82,025 ft) in just under four minutes. These flights set World time-to-climb records.

The same month NASA called for applicants to join the second class of astronauts. Young couldn't wait to apply. He fitted the

bill perfectly and got the job at the age of 31, being officially introduced to the press on 17 September 1962, along with eight others, after a short period in Fighter 143 Squadron in Miramar, California. On 3 October 1962, Young and his fellow eight class-two astronauts were on hand at Cape Canaveral for the launch of Wally Schirra in Mercury capsule Sigma 7.

Within 18 months, Young was assigned to be pilot of the first manned Gemini mission. He was named on 13 April 1964 and his performance at the press conference with command pilot Gus Grissom was such that he might just as well have been the invisible man. Gruff Gus wasn't much better and the two became known as the Taciturn Twins. The twins were totally overshadowed by Soviet cosmonaut Alexei Leonov's spacewalk five days before they blasted off beneath a Titan II intercontinental ballistic missile. The three-orbit flight of Gemini 3, nicknamed 'unsinkable Molly Brown' by the sinkable Grissom, was a modest success, despite becoming the first in which in-orbit manoeuvres had been performed. The spacecraft also carried an innovation, a computer, and was the first to be controlled from Houston rather than Cape Canaveral. The spacecraft overshot the splashdown zone by over 50 miles and the crew had a long wait. In the swell of the Atlantic, Grissom was sea sick. It got very hot in the cabin and the crew decided to remove their suits. Young and his commander thus made the traditional walk along the red carpet of the recovery ship, wearing long johns partly concealed beneath bathrobes.

Aboard Columbia *on flight 6.*

It was revealed after the flight that Young had smuggled a sandwich aboard Gemini and had offered it to a startled Grissom who, gamely had a bite, shedding a few crumbs into the weightless cockpit. The irrepressible joker Wally Schirra was responsible for this prank, having brought the sandwich in Cocoa Beach the evening before the launch and put it in Young's spacesuit pocket. The flak started flying. The contraband sandwich became known as the '$30 million sandwich', as Congress questioned this flouting of flight rules. As a result, Young had to be reprimanded. Schirra and Tom Stafford had served as Gemini 3 reserves and Grissom and Young repaid their service as back-ups on Gemini 6. This would have put

Young in line to command Gemini 9, but instead, like the inexperienced Neil Armstrong, who was placed straight into Gemini 8, Elliott See became Gemini 9 commander, only to be killed in an air crash later.

Young became Gemini 10 command pilot, replacing Edward White who had gone from Gemini 7 directly to Apollo, joining Grissom and Chaffee for the ill-fated first mission. With Mike Collins as his side-kick, Young had quite a job on his hands and the careful engineer inside him told him that Gemini 10 could present problems. Firstly, Young rendezvoused and docked with an Agena target rocket, using up more propellant than planned, threatening to curtail some later activities. For the first time, the engine on the Agena was restarted and propelled the duo to the then record height of 465 miles. Then the combination rendezvoused with Agena 8 and Collins made a spacewalk to retrieve samples from its side. He had to be called back inside because Young was concerned about the dwindling station-keeping propellant supply. During an earlier stand-up EVA, lithium hydroxide from the craft's life support system leaked into the astronauts' air supply, choking them and stinging their eyes. It was an eventful flight to say the least. Young then moved into the Apollo programme and his plans were almost immediately affected by the fire.

He was assigned to be back-up command module pilot of Apollo 2B, which, in 1967, was to be a flight involving the Apollo command and lunar modules launched separately on Saturn 1Bs. After the Apollo disaster, Young was reassigned to a similar position on what was to be the first manned Apollo flight, Apollo 7, thus making him elegible for Apollo 10, which at one time was thought of being capable of making the first lunar landing. During the Apollo 10 mission, launched in May 1969, Young became the first man to fly solo in lunar orbit and at some points of the lunar orbit phase, the loneliest man in the universe, out of direct touch with the Earth. By calling the Apollo 10 craft *Charlie Brown* and *Snoopy*, the astronauts set the tone for the mission which, although at times was decidedly risky, was quite lighthearted with Young joining the fun. Young then teamed up with Jack Swigert and Charlies Duke to be back-ups to Apollo 13. During this epic, heart stopping drama, Young provided vital support at Houston. Young was obviously married to the space business and not suprisingly, his marriage to Barbara suffered. Getting a divorce, Young quietly married psychology graduate Susy Feldman and the two set up home in an apartment across the road from the space centre. Naturally, Apollo 16 was to follow; Young's eighth assignment in ten years with NASA.

After backing-up Apollo 17, forging a path to the Space Shuttle, Young was immediately put in charge of the Shuttle astronaut branch. He became chief of the astronauts in 1974 and in March 1978 was named commander of the first Shuttle mission which was due to take place in 1979. Bob Crippen and Young became the most intensely mission-trained crew ever because the flight was delayed over and over again. The Shuttle had earned the description white elephant by the time it took to the skies in 1981. In addition, the USA had been through a technologically traumatic period, the Five Mile Island nuclear alert and the Chicago DC10 disaster included. There were many at the Kennedy Space Centre who felt that Shuttle *Columbia* could be added to the list. Young and Crippen were the first people to make a spaceflight in the maiden flight of a space vehicle type. The mission was an heroic one. On 10 April 1981 the launch was scrubbed at the last minute due to a computer fault. The quietly confident astronaut, now turned 51, was unruffled by it all, which is more than can be said of the popular press which made an unholy meal of the cancellation, calling it a 'fiasco'.

Young in 1986, with little prospect of another flight.

Twenty years to the day that Yuri Gagarin became the first man in the space, Young took Crippen aloft to become the

102nd, while he himself was to become the first to make five spaceflights. At T-10sec the ignition sequence started, the Shuttle main engines were alight by T-6sec and up to full thrust at zero when two simultaneous flashes signalled solid rocket booster ignition and a cataclysmic blast-off. Within seconds, *Columbia* was startling unprepared observers by arcing over on its back, leaving a trail of dense golden flame – and an extraordinary din. STS 1 was a brilliant mission, epitomised by the matter of fact image of Young, wearing half-moon glasses perched on the end of his nose, reading a checklist and looking for all the world like an airliner pilot. The landing of *Columbia* at the end of the mission two days later, captured the imagination of the world. Young had never been in such high spirits of jubilation before. He and Crippen were honoured by President Reagan, who then cut the space budget. Young remarked, "politicians are a strange bunch of critters".

Still, Young served on, returning to the chief of the astronauts' desk, vacated by Al Bean who had kept it warm for Young during his training for STS 1, and being named commander of yet another mission, his tenth assignment in 20 years with NASA during which he had clocked up 10,000 hours in specific mission simulations. This was also on *Columbia* and the ninth mission in the Shuttle programme overall. Launched in late November 1983, STS 9 became the longest mission, lasting ten days. Carrying the Spacelab laboratory with a record crew of six, with Young making history as the first to make six spaceflights, it was a little longer due to a computer fault which "turned my legs to jelly," remarked Young afterwards. The firing of the re-entry engines was suddenly halted with a lurch and crunch. One of the suite of five computers had malfunctioned and it took Young hours of troubleshooting before he brought *Columbia* home to roost. Even then, the troubles were not over because as it taxied to a stop, the tail of *Columbia* was clearly on fire. Propellant had leaked.

Young became intimately involved in all Shuttle missions. It was usually he who would fly the weather aircraft round and round the sky before a launch to ensure that conditions were suitable. Flying over the launch pad during the actual launch of Shuttle, Young would often come away with some fine photos, the one showing the ascent of STS 61A in October 1985 being particularly good. He didn't get a picture of an earlier launch, 51D in April 1985, however, because it took off against his wishes, in thick drizzle-filled cloud. The Shuttle was straining on the pad and heading for a launch scrub, mainly because Young flying over head, was reporting water on his wind shield. But Senator Jake Garn was aboard this Shuttle and he had helped to clear NASA's budget. The space agency's administrator strode into the firing room and the go for launch was given. STS 51D wasn't the only risky launch; 51I in August 1985 was fired through a hole in heavy thunder showers.

These decisions rankled the ever-cautious Young, who became concerned that safety was not being placed first on the list of priorities. He even fired off a memo or two on the subject. Commercial and political pressures were at the top of the list of Shuttle priorities and the inevitable happened. "I was hoping against hope it wasn't going to happen but it did," said Young afterwards. He fired off another memo. His memos were made public and Young became the centre of controversy. What he said was not questioned; it was the way he was saying it that was. Some astronauts were critical of his ineffectiveness in getting the issue sorted out before *Challenger*. The astronaut office was not in a strong position before *Challenger*, even crew commanders were not involved in vital pre-launch meetings. After *Challenger*, the office became all-powerful and an important influence on the Shuttle rehabilitation programme. But Young was not really part of it anymore. But if he left NASA, what else could he do? Join the Soviet cosmonaut team, probably.

Charles Duke – the Convert

On selection in 1966.

56th person in space
11 day 1 hr 51 min space experience
20 hr 8 min walking/driving on moon

CHARLIE DUKE WALKED IN THE VALLEY of Descartes, on top of a mountain range on the Moon. He collected samples of lunar rock of all ages. After the flight, he said that he believed that God made the Moon, instantly during The Creation. One minute it wasn't there, the next it was. Charlie didn't come to this conclusion at the post-flight press conference but many years later, after his life had been transformed by The Bible and by Jesus Christ. Today he tells everyone that he may have walked on the Moon but today he walks with the Son.

Charles Moss Duke Jr was one of twin boys born on 3 October 1935 in Charlotte, North Carolina. He was interested in planes as a boy, watching them soar over his backyard at Lancaster, South Carolina. "I've always been ambitious and I suppose I aimed for the Moon ever since then." Perhaps it was really when he graduated from US Naval Academy the year that Sputnik 1 was launched that he was fired-up. Top of his class at graduating from the Admiral Farragut Academy in St Petersberg, Florida, before this, he graduated with a degree in naval sciences from the Academy with honours. "Due to chronic sea sickness, I just couldn't become an ocean going Navy man, so I transferred to the Air Force,"[1] he said. He learned to fly at Spence Air Force Base, Georgia and followed this up at Webb Air Force Base in Texas from where he graduated with distinction in 1958. He was again a distinguished graduate from Moody Air Force base, Georgia where he completed advanced training in F-86L aircraft. He was then posted to the 526th Interceptor Squadron at

Official Apollo 16 portrait.

Ramstein Air Base, Germany where he served an an interceptor pilot for three years.

By this time, NASA had chosen three teams of astronauts. NASA wanted pilots with degrees, so, in 1964, Duke completed a course at the Massachusetts Institute of Technology, graduating with degrees both in aeronautics and astronautics. "Being success oriented, I set my sights on higher and higher levels of achievement." He also got married to Dorothy Meade Claiborne of Atlanta, Georgia, or just plain 'Dotty'. A year later his ambitious path to the Moon led him to the famed Edwards Air Force Base and the Aerospace Research Pilot School, from where he graduated in September 1965 just after Gemini 5 had landed and in the year when his first son, Charles Moss, was born. He stayed on at Edwards as an instructor teaching control systems and flying F-104, F-101 and T-33 aircraft and was still there when he applied to join the fifth astronaut group in April 1966. "One morning I read a full page ad in *The Los Angeles Times*. Overjoyed at the opportunity to again climb the ladder in my career, I applied."[2] At 5ft 11½in tall, the wiry 155 pound pilot was lucky to get in. If he hadn't made the Nasa group, he would have flown the X-15 rocket plane and made it into space another way.

"My new job brought me instant success – name in the paper, invitations to the nicest parties," where Duke would enjoy chatting up the ladies, much to the chagrin of Dotty, who was beginning to get very depressed and lonely. "That old ego swelled bigger and bigger. I liked this new job and all it brought with it. I worked hard. I wanted that top flight – a trip to the Moon. The next years were dedicated to long hours and work and a full social life."[3] His marriage was on the moonrocks. "I turned my marriage and family over to Dotty. There was not much time for or interest in this area of my life." Within a year of training and the birth of his second son, Thomas Claiborne Duke was catching the eye of the potential Apollo commanders but did not make the support crews of the first Apollo missions. Dotty, meanwhile, was beginning to feel "heaviness and loneliness".

As the support crew member of Apollo 10 and cap com during important phases of the mission, Duke's Carolina drawl and humour captured the attention of a listening world who probaly didn't know who they were listening to. Apollo 10 commander Tom

Training with Young.

Stafford knew who he had listened to during the critical moments of the mission and Neil Armstrong, commander of Apollo 11, wanted to hear him too, asking Duke personally to be on the support crew of his mission and cap com during the first moon landing. Thus Duke went into the history books with his breathless "Roger Tranquillity, we copy you on the ground. You've got a bunch of people about to turn blue. We're breathing again, thanks a lot".

Assigned to back up Apollo 13, shortly afterwards Duke was thus in line for Apollo 16 – until the budget axe fell, threatening him with the replacement of geologist Jack Schmitt. Duke needed luck. He was named lunar module pilot for the mission on 3 March 1971, after Apollo 14, and his landing site at Descartes was confirmed officially on 17 June the same year. Apollo 16 was originally scheduled to take off on 17 March 1972, at the time Duke chose to land up in Patrick Air Force base hospital with pneumonia. Fortunately for him, the launch date was put back to 16 April by suit fitting, lunar module battery and docking ring jettisoning system problems. His back-up was Ed Mitchell. Would Mitchell have flown on 17 March becoming the first person to make successive spaceflights and the first to make two Moon landings? Surely, the mission would have been delayed a month anyway, for Duke, wouldn't it? Duke had already achieved fame with his ill-health, causing the Apollo 13 scare by catching German measles to which Apollo 13 command module pilot had no immunity. This cost Mattingly the flight, which he may afterwards have considered fortunate in the circumstances. He ended up on Apollo 16 too.

In the Apollo 16 press kit, Duke listed his hobbies as hunting, fishing, playing golf and reading, thus proving to be one of the few, if not the only, astronaut to list reading as a hobby. The others probably felt that they did enough of that already with briefing books. Like a true Carolinan, Duke asked for and got dehydrated grits on his Apollo breakfast menu.

"Tall, scarecrow skinny, with wide-open smiling face and brown eyes that almost pop, like the eyes of a kid in a Saturday matinee,"[4] is how Duke has been described. The easy going, drawling Carolinan was considered an ideal foil for the taciturn commander John Young. In fact, Young was by no means taciturn during the flight and his and Duke's banter on the Moon made the mission a wow with the space fans. "Yahoo, this is so great, you can't believe it." Drilling into the moon, Duke said, "Look at that beauty go." When the drill stuck, he said, "look at that beauty stop." He even did a passable impression of W. C. Fields while helping Young to load some equipment on the rover.

The simple fact of the matter is that Duke nearly killed himself on the Moon. Going for the lunar high jump record, he squatted and leaped. He rose magestically five feet into the 'air' but the weight of his portable life support system backpack pulled him backwards, ever so slowly. Ever so quickly, Duke realised that he had 'bought the farm' on the Moon. The PLSS would smash on impact and he would die a swift death. As he descended to the surface on his back, directly above him he saw the Earth in the sky, like the Sun at noon. He landed badly but his PLSS and he himself survived. He later described what could have been his last mortal sight, the Earth, as "this beautiful jewel hanging up in the blackness of space." [5] The fun continued and Duke made it home.

Walking the highlands of Descartes.

He would say that the trip did not effect him spiritually but he had a feeling of being at home on the Moon. It wasn't like Buck Rogers, he said but "it was beautiful, crystal clear, serene. One did not have the impression of a hostile environment but of beauty. I walked in wonder on the Moon, breathless in admiration, in that incredible still, crater pocked awesome environment, coloured in shades of grey. It was just the way it had been created – pure, unspoiled, untouched. I was proud to be one of the few men to have such an experience"[6]. Despite this, however, outwardly, he said, he was just the same easy-going sort of guy he was before he went to Descartes. He was never recognized 'in the street'. His kids brought him down to Earth and "didn't let you get all inflated". But, like most the men who went to the Moon, the view of Earth would have a lasting impression. His "traditional parochial outlook was vastly stretched". He was still 100 per cent American, he said, but he felt insignificant holding his hand up and covering the Earth, when he was at Descartes. "Now I see Earth and mankind more as a whole than as individual races and religions and nations."[7].

Instead of retreating to his family at last, the dedicated astronaut went back into training for Apollo 17, as back-up to Jack Schmitt. "After the flight, our marriage didn't improve and Dotty became very depressed, even suicidal. She had put our relationship first in her life and it was failing"[8]. After Apollo 17, Duke was assigned to work on the

Space Shuttle programme as technical assistant for orbiter integration. He later assumed additional duties as department manager for advanced planning for Shuttle. He was clearly a potential future Shuttle commander. But things weren't right with him. "Frustration and boredom set in. My job had been the most important thing in my life and I had reached my goal of going to the Moon. I asked myself, 'what am I going to do now'"[9].

Wanting a completely new challenge and to make pots of money and to be a good provider for his family, Duke opened a Coors beer distributorship in San Antonio, Texas. He left Nasa at the end of 1975, having waited two years for the distributorship to become available[10]. "My new goal was to be a successful businessman and to make a million dollars," Duke explained in his own evangelical leaflet. "We moved to New Braunfels, Texas, where I completely immersed myself in my new job, working long hours. But two years later, I decided that I was in the wrong business, not aware that I hadn't dug to the root of my problem."

Even though Duke had been into the heavens, he hadn't found God on the Moon. In fact, he hadn't found God in Church although he attended reguarly, mainly to keep Dotty company. "I read the scriptures without believing a word." Duke felt that Jesus was a great teacher, like Buddha. After leaving NASA Duke had also noticed a change in Dotty, who told Charlie that she felt that her prayers were being answered and that she had a sense of purpose and peace in her life that whe had never had before. "She began to love and accept me in a different way".

A month after selling the business, Dotty asked Charlie to go to a two-day Bible study at a tennis ranch nearby. "We started at Genesis and headed for Revelations. As I sat in an easy chair with a cup of coffee in my hand and the Bible in my lap, the scales suddenly fell from my eyes, I saw that God had loved Charlie Duke from the time He created the world. I saw how man had turned away from God, who constantly gave the same message: 'Turn to Me, and I will be your God and bless you'. All of a sudden I realised that I was faced with the most important question I had ever faced. In my heart Jesus was saying to me "Who do you say that I am?" The only choices I could see were that Jesus was either the Son of God or a big liar. As Dotty and I drove home, I looked at her and said, "Love, there's no doubt in my mind, Jesus Christ is the Son of God." And Charlie Duke was born again."

Duke in 1988.

Charlie felt no blast off to eternity but "I knew from God's word that by believing He came into my heart. I got hungry for the word of God from that moment on." Duke started praying, "Lord teach me the truth written in the Bible. Help me to love my wife. I want to be a good husband and father." Duke says that God did just that. He put love in Charlies heart for Dotty. He healed their marriage. "We have put Jesus first in our lives and the closer we get to Jesus the closer we grow to each other." As the months passed, Charlie desired to know Jesus more. One night, he awoke and felt a strong presence of the Lord in his room. He knelt down and "surrendered my whole life to Him. Right then I was filled with the Holy Spirit and received the power to tell others about Jesus and to live fully for him". Duke and his wife prayed for the sick and say they have seen God perform miracles. "We have seen blind eyes opened, the deaf hear, cancers healed, lives changed."

Now a world wide traveller and lay preacher, as well as an owner, president, director and secretary of several companies, Duke has never known such excitement – "a life filled with the love, peace, joy and the power of God." Duke rode a fantastic adventure on Apollo 16 and used to say that one could live 10,000 years and never have an experience like walking on the moon. "But the excitement and satisfaction of that walk doesn't begin to compare with my walk with Jesus, a walk that lasts forever – a walk with the Son."

Gene Cernan – the Ambassador

On selection in 1963.

27th person in space
12th person to make two spaceflights
9th person to make three spaceflights
23 day 14 hr 16 min space experience
22 hr 4 min walking/driving on moon

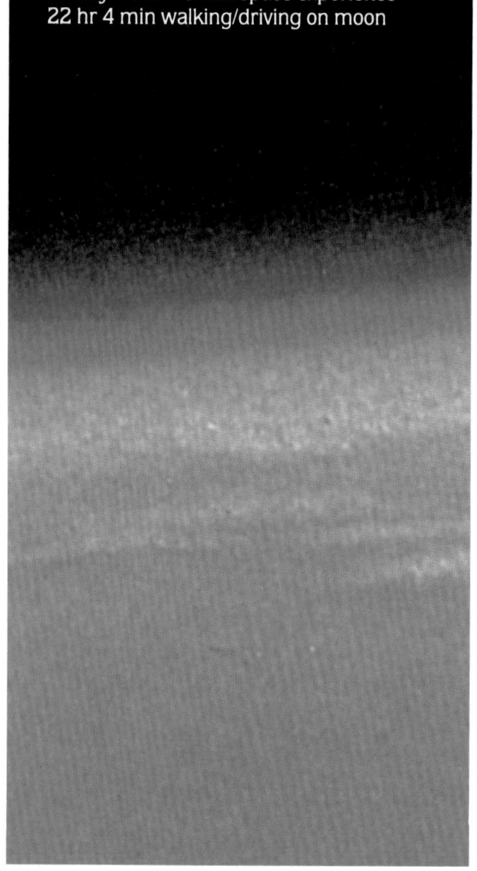

OF THE 14 NASA ASTRONAUTS CHOSEN IN October 1963, Gene Cernan was considered as the least likely to succeed in the space business, according to one of his peers, Walter Cunningham, in his autobiography. Ironically, it was Cunningham who had the hard time and Cernan a lot of luck. Like the time he was flying over the Banana River near the Kennedy Space Centre, before the launch of Apollo 14 for which he was back-up commander. He was said to have been distracted by certain activities taking place in the secluded ground beneath him. His helicopter crashed into the river and he escaped by swimming out of the cockpit underwater. There were some who questioned whether he should take command of Apollo 17 after this. He did and his performance towards the end of the lunar exploration illustrated his ability to communicate with loquatious geniality, hidden beneath a confident, bullish personality. It wasn't surprising to see him becoming an ambassador for space both in support of NASA and his own business, as president of The Cernan Corporation, a flourishing concern in Houston.

Eugene Andrew Cernan was born in Chicago, Illinois on 14 March 1934. His parents, Andrew George Cernan and the former Rose Ann Cihlar, were first-generation Czechs and lived in a blue collar suburb of the city. Cernan was an excellent athlete and after graduating from Proviso Township High School, turned down a football scholarship to graduate with a degree in electrical engineering at Purdue University in Lafayette, Indiana as part of a Naval Reserve Officers Training Programme. After graduation, Cernan, with his eyes on flying jets from aircraft carriers, became an ensign in the Naval reserve, reporting for duty on the *USS Saipan* on 25 June 1956. In October of that year he entered Naval Air Basic Training Command at Pensacola in Florida, moving to Naval Air Advanced Training Command in Memphis, Tennessee in September 1957. Lt Cernan received his wings in November 1957 at a time when the US space prgramme was at its lowest ebb, having experienced the Vangaurd 'flopnik' disaster in the shadow of the Sputniks. He received advanced training with Attack Squadron 202 back at Pensacola until February 1958, before being assigned to Attack Squadron 126 and then to Attack Squadron 113. On 5 May, 1961, the day

Inside Gemini 9 (left), with Tom Stafford, in 1966.

after Alan Shepard became the first American in space, Cernan married Barbara Jean Atchley at the Naval Air Station Mirama in California. The following month, Cernan entered the US Naval Postgraduate School in Monterey, California and by the time he had gained his masters degree in aeronautical engineering in January 1964, he had already been named as a member of the third group of NASA astronauts in October 1963, the year of the birth of his only child, Teresa Dawn or Tracey. Cernan did't bask in the

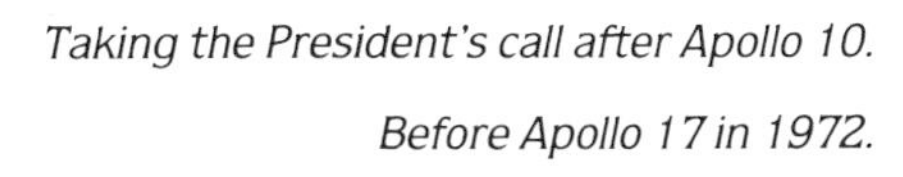

Taking the President's call after Apollo 10.

Before Apollo 17 in 1972.

glory of it. Working in his front garden in Houston, he was confronted with the driver of a car who asked him "where do the astronauts live"? Cernan told him with a cheeky smile, "I think there are some down there at the end of the block".

Within two years, Cernan, by then a Lieutenant Commander, was down to back-up Charlie Bassett on Gemini 9 and would have then taken the final Gemini into orbit in late 1966, making a late entry into Apollo, perhaps to the detriment to his prospects of an early mission. Like so many of the astronauts, Cernan's prospects were improved by another's tragedy. Bassett, who was to make a spacewalk, and his commander Elliott See, were killed in the St Louis plane crash. Cernan and Tom Stafford, the back-up command pilot of Gemini 9 were flying behind the prime crew but did not see the crash because of the thick fog that caused Bassett's ill fated approach. Landing safely, the astronauts learned of the accident that placed them immediately on the prime crew of Gemini 9, just three months away.

Tom Stafford had already experienced a launch pad abort on Gemini 6 and added to his trips to and from the launch pad without lifting off for Gemini 9 which became a jinxed mission before it took off. First, its Agena target rocket exploded and the astronauts had to return to their quarters. Next, just two minutes to blast-off, with a replacement target rocket in orbit, the Titan booster misbehaved and again, it was back to the crew quarters. Stafford seemed quite distraught but Cernan, described as a deeply religious Roman Catholic, seemed serene and comforting to his commander. Finally, Cernan got flying. A relieved Stafford rendezvoused with the target but could not dock with it because the shrouds had not come off.

Cernan, the youngest American astronaut to fly and still today the youngest American male in space, made a spacewalk during which he was to don a prototype astronaut manoeuvring unit and perform pirouettes in space. The spacewalk became dangerous when the sweating, over heating Cernan's faceplate became fogged. "I can see where my nose is but not where my eyeballs are," he said. Called inside, before having a chance to don the unit, Cernan could still find time to crack a joke or two. Gemini then performed its best parting shot, the most accurate landing until the Shuttles, just yards from the deck of the carrier.

Cernan then went into the most dead-end assignment imaginable, backing up Gemini 12, the final flight, as pilot – not even command pilot – while others went on to seemingly greater things. With Cernan was Gordon Cooper, whose lack of success in getting an Apollo off the ground was to be quite spectacular. But Cernan was involved in the great Apollo crew shuffle and came out of the pack with a good hand, eventually. He was back-up lunar pilot of Apollo 2B which was to be the first flight of all three Apollo modules.

At Taurus Littrow, the last man to leave the Moon.

After the Apollo fire, Cernan, with Stafford and Young backed-up Apollo 7 and took Apollo 10 to the Moon on the riskiest space mission to date. The command module was nicknamed *Charlie Brown* and the lunar module, *Snoopy*, characterising the nature of the crew, despite the seriousness of their mission, which if all had gone to plan, would have made the first manned landing, making Cernan the second man on the Moon. Instead, Apollo 10 was to be a complete dress rehearsal of the landing, 'landing' nine miles above the Moon. Things went well. The extremely excited Cernan, was a real chatterbox. "Aw, Charlie, we just saw Earthrise and its got to be magnificent," he said as *Snoopy* swooped towards the Moon.

Suddenly, when the ascent stage separated from the descent stage, simulating take off, it went into a spin. If this had happened during a real lunar take off, the first men on the Moon would have been the first to die. Things were pretty serious nine miles above the Moon. As Stafford struggled to regain control, Cernan exclaimed, "Sonofa bitch, there's something wrong with the gyro.... I don't know what the hell that was babe.... I'll tell you there was a moment there Tom.... [When Cernan had thought this was the end] .... Let's worry about it when we make this burn.... OK, Charlie, I think we got our marbles.... We're sure coming down to that ground, I'll tell you.... I don't know what happened there but I hope we never find it again." Cernan calmed down on the return journey even announcing that he felt 'horny', which was rare thing for an astronaut to report publicly.

After Apollo 11, Mike Collins, its command module pilot, was told that he could be back-up commander of Apollo 14, probably taking Apollo 17 to the Moon. Instead, he chose to resign, having had enough of the space business and having been part of the most historic voyage in history. Cernan was appointed to the task, taking on the rookies Ron Evans, his old Navy pal, and Joe Engle a dashing former X-15 rocket plane pilot and recipient of astronaut wings for flying it over 50 miles high three times. It was obvious to observers that this crew would be rotated naturally to become the prime crew for

Apollo 17. But there were complications; Cernan crashed his helicopter just before the Apollo 14 mission and Jack Schmitt had lost Apollo 18 due to budget cuts. There was a suggestion that Apollo 17 would be taken over by the 18 crew, with Richard Gordon and Vance Brand. But in the end, Cernan survived, with Evans but without Engle who was replaced by Schmitt.

The grey haired 38-year-old Cernan was the commander but Schmitt was the scientist. Both were under pressure, Cernan as the scientist and Schmitt as the pilot astronaut. The difference between to the two was often summed up as 'The astronaut and scientist' or 'The scientist and veteran'. Like all the astronauts and pilots in general, Cernan was confident. He told *Spaceflight News*, "I just knew that not only could I do it but I could do it better than it had ever been done before and that's the attitude a good aviator has. There's no question of a screw-up. You have to have that attitude because if you're worried about all those other things you shouldn't have gone in the first place."

Chief of the Cernan Corporation.

Cernan was also immensley patriotic. "For me it wasn't that man first stepped on the Moon," he said of Apollo 11, "it was than an American was planting the American flag for all to see." Cernan called his landing site Camelot, in tribute to the late President Kennedy, whose White House were often dubbed Camelot by the press and whose favourite musical was about King Arthur's castle. Just before leaving Camelot, Cernan's last words on the Moon were, "OK let's get this mother out of here".

According to an article in *TV Guide*, the Catholic Cernan had found God before he went to the Moon but the trip made the realization of God more powerful. What he saw was "almost too beautiful to grasp. There was too much logic, too much purpose, it was just too beautiful to have happened by accident. It doesn't matter how you choose to worship God, he has to exist to have created what I was privileged to have seen".

After Apollo 17, Cernan supported the Apollo Soyuz Test Project flight in 1975 and on 1 July 1976, he resigned from both NASA and the US Navy, with the rank of Captain, to join Coral Petroleum Inc of Houston to be, unashamedly a marketing-cum-PR man, using his fame. He was a fine ambassador, sharing his experiences and supporting the US space programme vehemently. He is sad at the low state of America's space programme. It seems such a waste to him that after what Amercia proved it could do, there's nothing following. Political support is not matching public awareness of the possibilities and support for projects such as Space Station and, later, flights to the Moon and Mars. Cernan fears that America will take second place in space, if it hasn't already.

In 1981, Cernan formed his own corporation, an aerospace and energy management consultancy and was co-anchor on several ABC TV Shuttle launch specials. He is now divorced from Barbara. He walked on the Moon and he's proud of it. All these stories about some of the moonmen not be affected by the flight was rubbish, he said. "You can't go there and come back without having some memories and feelings indelibly etched in your mind. What the Moon did was give all the guys a platform from which to further their studies and exploits," he told the *Observer Magazine*. But of those that were overtly affected he said, "The truth is that Buzz Aldrin was what he was before he went to the Moon, Ed Mitchell was an ESP freak before he went and Jim Irwin would have probably ended up an evangelist." Cernan? He went and is making every moment of it. Cernan told the Florida newspaper, *Today*, "What could I do to match living and walking and driving on the Moon for three days? There are only two people who have been to the Moon (lunar orbit) twice, and I'm one of them.

Twelve guys walked on the Moon and I happen to be the last guy to have left my footprints there. I'm the luckiest guy in the world,"[6] he says.

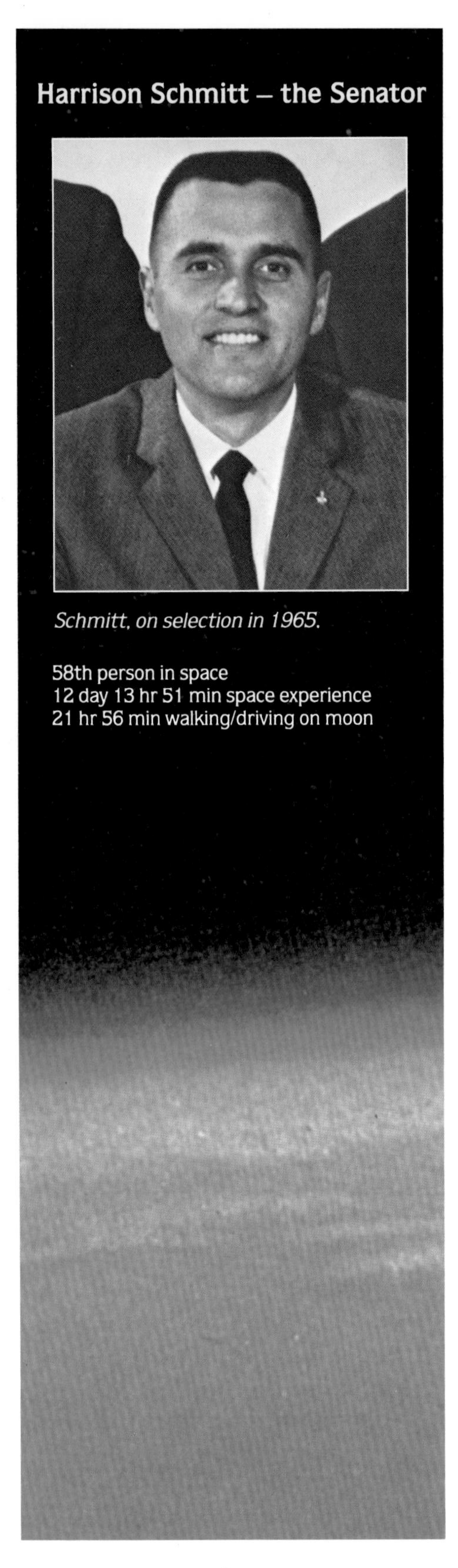

Harrison Schmitt – the Senator

Schmitt, on selection in 1965.

58th person in space
12 day 13 hr 51 min space experience
21 hr 56 min walking/driving on moon

STUDYING FOR HIS DOCTORATE IN geology at Harvard University in 1963-64, Harrison H. Schmitt had a long-term game plan, his eyes fixed firmly on a target which he expected to hit in the bullseye. But that wasn't the Moon. It was the US Senate, to which 'Jack' was eventually elected in 1976. It would have been sooner had the Moon not got in the way – but it sure helped a bit, too.

Schmitt was born in Santa Rica, New Mexico on 3 July 1935, the son of a mining gelogist who the young Jack would soon want to emulate, visiting mining camps, exploring Indian reservations and rock hunting in the lunar-like landscapes of the American southwest. Schmitt graduated from Western High School, Silver City, New Mexico in 1953. He then studied Earth sciences at the California Institute of Technology, graduating with a degree in 1957, the year of Sputnik; not that Schmitt would have taken much notice. As a Fulbright Fellow, he studied at the University of Oslo in Norway in 1957-58, while at the same time working for the US Geological Survey in western Norway. With the Geological Survey until 1961, when he joined Harvard University to teach, Schmitt also worked at sites in southeastern Alaska, his home state New Mexico and in Montana. He taught a course in ore deposits at Harvard, before taking his PhD in geology, graduating in 1964, to join the Geological Survey's Astrophysical Centre at Flagstaff, Arizona.

It was here that Schmitt finally caught the space bug, specialising in photographic and telescopic mapping of the Moon, using the first close-up pictures taken by the US Ranger spacecraft. These photos were long awaited because many Rangers failed before, at last, in July 1964, Ranger 7 plummeted towards the northern rim of the Sea of Clouds, its six TV cameras taking 4,315 pictures before the final picture was exposed a second before impact. Schmitt also had another job, training NASA astronauts in geology and taking them on field trips to simulate lunar exploration. Some of these astronauts would never get to the Moon. The fact that Schmitt might, had not even occurred to him until suddenly NASA announced it was recruiting not test pilots but scientists as astronauts. Within a year, Schmitt was at Houston with five other astroscientists. One soon resigned because of divorce, something which astronauts just could not do. As the only geologist in the group, Schmitt was almost assured of a moon flight, provided the first missions were successful – and that the budget held out.

They may not have been test pilots but the astroscientists had to learn to fly and were packed off to Air Force flight school. Schmitt's 53-week stint at Williams Air Force Base was a spectacular success, with the geologist coming second in his class. In addition to basic astronaut training, Schmitt was available on-tap to NASA providing Apollo flight crews with detailed instruction in lunar navigation, geology and feature recognition. He also helped to design various Moon tools and scoops. Schmitt was clearly doing all the right things; he couldn't even get divorced because he wasn't married. It was just a matter of time before he managed to get on a flight crew. But it was fellow astroscientist Ed Gibson who got closest first. Gibson, an electrical engineer, was taking helicopter lessons while in the support of the Apollo 12 flight, the inference being clearly an association with a lunar landing. But at the same time, the budget axe was being wielded heftily and one by one, Apollo trees were lopped down unceremoniously. One astroscientist, Curt Michel, had, meanwhile, objected to science taking second place in NASA and left. Two down four to go. This was extremely convenient to NASA, meaning that three could be assigned to the three planned Skylab missions and one to an Apollo flight. It had to be Schmitt.

So, in late 1969, Jack Schmitt was assigned to be back-up lunar module pilot of Apollo 15, making him elegible to make a Moon landing on Apollo 18. The Apollo 13 'magnificent failure' delayed Apollo 15 from late 1970 into 1971 and then the budget axe came out again. Although three Apollo flights were to be cancelled, they did not specifically include Apollo 18 by name. Schmitt was theoretically safe but in the end NASA retained Apollos 15, 16 and 17 as they existed and cancelled 18, 19 and 20. Schmitt and his commander Dick Gordon had lost the Moon. Apollo 14's back-up, crew Gene Cernan, Ron Evans and Joe Engle, were eligible for Apollo 17 and it was this crew list that was submitted to NASA headquarters for rubber stamping in late 1971. The list came back to Slayton. It read Cernan, Evans, Schmitt. The scientific community had won at last; a scientist was going to the Moon. Engle, who must have seen the writing on the wall long before this, stayed on at NASA, making the second Suttle test flight ten years later.

Schmitt immersed himself in Apollo 17 training manuals and checklists and had to work to become part of the crew which was new to him. So dedicated was he that it was reported that when spaceworkers took him to lunch at a topless bar at Cocoa Beach while he was preparing for the flight, he spent all the time reading his manuals and didn't notice the exquisitely proportioned pair of boobs being dangled before him. Schmitt did have time for female company and had formed a firm friendship with the widow of astronaut C.C.Williams who was robbed of the Moon by an air crash in 1967. And it was Beth who was waiting to embrace him when he returned from the Moon in December 1972.

Schmitt had performed well on the Moon but no better than his commander Cernan, who had ironically been his pupil in geology. That isn't a reflection on Schmitt's capabilities but testimony to the limited time Schmitt was given on the Moon, something that perhaps epitomizes the Apollo programme.

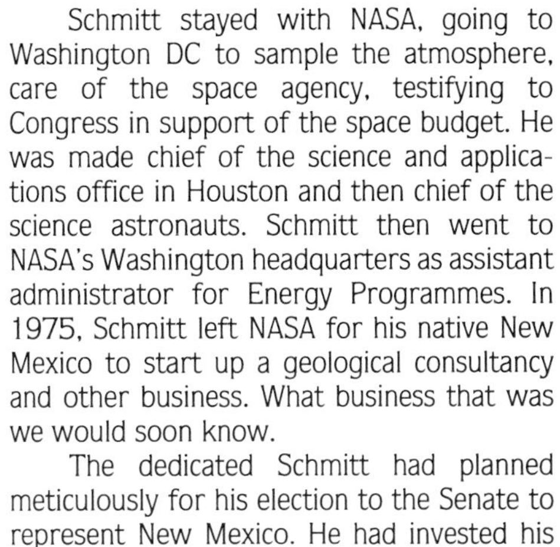

Schmitt stayed with NASA, going to Washington DC to sample the atmosphere, care of the space agency, testifying to Congress in support of the space budget. He was made chief of the science and applications office in Houston and then chief of the science astronauts. Schmitt then went to NASA's Washington headquarters as assistant administrator for Energy Programmes. In 1975, Schmitt left NASA for his native New Mexico to start up a geological consultancy and other business. What business that was we would soon know.

The dedicated Schmitt had planned meticulously for his election to the Senate to represent New Mexico. He had invested his share on the *Time Life* money since 1965 to pay for his campaign funds. "One of the things that I learned from the space programme is to how to schedule a plan and stick to it"[1]. He waged battled against incumbent Democrat senator Joseph Montoya. Of his republican opponent, who was described in the election papers as a consultant geologist, Montoya was heard to comment, "In those days I used to support the idea to take a man to the Moon; however, I should attach to that the condition to leave him there". Schmitt's blunt talk and policies, and astute planning, won him, in 1976, the seat in the Senate, the first time in New Mexico's history that an incumbent US Senator had been turfed out. The voting figures were: Montoya 174,308 votes (43 per cent), Schmitt 231,515 votes (57 per cent). So, Senator Schmitt went to Washington. Ex-astronaut John Glenn was already there as Senator for Ohio. TV commentator Walter Cronkite remarked that astronauts consituted only two ten-millionths of the US population, yet two were in the Senate. He humourously called this, "representation out of all proportion".

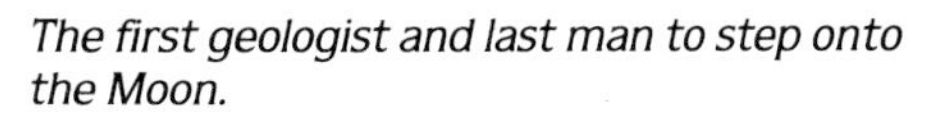

The first geologist and last man to step onto the Moon.

Jack Schmitt served his State through the 95th, 96th and 97th Congresses and was, naturally, a leading spokesman on space affairs, breakfasting with the first Shuttle crew before their blast-off in April 1981. He served as chairman on various committees and subcommittees, covering a whole gamut of subjects from education to rural development. Schmitt also introduced a space policy bill whose goals echoed those of the National Commission on Space published after the *Challenger* disaster. The NCS's policy was equally as unsuccessful as Schmitt's earlier proposal. Schmitt was the chairman of the Commerce, Science and Transportation subcommittee on science, technology and space which tended to follow his lead on major decisions effecting NASA. The Senator was, however, also regarded as a loner who had little actual impact on the Senate, according to some observers. In the 1982 election, his Democrat opponent Jeff Bingaman made this his campaign theme. Schmitt's unconcerned reaction cost him votes and he lost by eight per cent. As he left Congress, another astronaut was entering it. Jack Swigert had just been elected a republican representive for Colorado. Sadly, he died of cancer shortly after.

After receiving an accolade in the Senate from fellow senator Moynihan, who cited Schmitt's technoligcal expertise, devotion to detail and conscientiousness, he re-entered private business in 1983, picking up a number of directorships. In October 1984, he was strongly considered to replace James Beggs as NASA's administrator. After Beggs finally left, falsely accused of fraudulent activities with his previous employer General Dynamics and in the wake of the *Challenger* disaster, many observers thought that the post was Schmitt's for the taking. But President Reagan turned to former NASA administrator, James Fletcher. If Schmitt was offered the job he probably turned it down, such were the circumstances. When a new administrator is chosen by a new President early in 1989, Schmitt will no doubt be considered again. Meanwhile, he beavers away and apparently commands fees of $4,000 a day, plus expenses judging by his conditions for an interview with me in preparation for this book. To be fair, Schmitt was writing a book of his own to commemorate the 20th anniversary of the Moon landings. Hope it sells well.

Schmitt (left) trains for Apollo 17 with Cernan.

At Taurus Littrow.

Senator Schmitt. (from H. Schmitt)

The Never Flights

THE SUMMER 1973; on THE FOURTH anniversary of the landing of Apollo 11. The Scene: Apollo 20 lunar module *Intrepid 2* provides a spectacular finale to the programme, touching down inside the giant crater Copernicus. The commander is Pete Conrad, the first man to make two Moon landings. His LMP is Jack Lousma. Orbiting above is Paul Weitz.

Had budget cuts not axed the last three moon landings, such a mission *would* have taken place, the exact place of landing being the only matter in doubt. There were flights that never were at the begining of the Apollo programme too, during which the fortunes of many an astronaut were decided.

The first manned Apollo mission, Apollo 1, commanded by Gus Grissom, Edward White and Roger Chaffee and backed up by James McDivitt, David Scott and Rusty Schweickart, was to have been followed by an almost identical flight, Apollo 2 (sometimes referred to as 2A). Inside the command module were to have been Wally Schirra, Donn Eisele and Walter Cunningham. It has been suggested that Schirra was not happy at being assigned to such an ordinary mission and was only placated when told that he was standing-in temporarily for Deke Slayton who would actually command the flight. Slayton was the chief of the astronauts but also a working astronaut himself until he was cruelly robbed of a Mercury orbital flight in 1962 by a heart flutter. Although he missed Gemini, Slayton felt he had a chance to convince the doctors that he could fly Apollo 2. He failed and only made it into space as a 51-year-old third pilot on the Apollo-Soyuz US-Soviet space extravaganza in 1975. The back-ups for Apollo 2 were to be Frank Borman, Tom Stafford and Michael Collins, the latter of whom had obviously been identified as a future lunar module pilot.

Apollo 2 was cancelled and Schirra and company were transferred to Apollo 1 as back-ups. This freed McDivitt, Scott and Schweickart to be assigned Apollo AS/205-208, sometimes referred to as Apollo 2B. A Saturn 1B would launch the astronauts in an Apollo command and service module and another would launch a lunar module. McDivitt would dock with the lunar module and perform limited low Earth orbit operations. The back-ups were Tom Stafford, elevated to commander status, John Young and Eugene Cernan, Stafford's former side-kick.

Apollo 205 would be followed by Apollo 503 or Apollo 3 as it is sometimes refered. This would be the first manned launch on a Saturn to perform full 'lunar' operations within the confines of an elliptical Earth orbit, with an apogee of about 4,000 miles. The commander would be Frank Borman. Michael Collins was elevated to command module pilot and the lunar module pilot would be William Anders. The back-ups were Pete

Conrad, Richard Gordon and, until he perished in an air crash, Clifton 'C.C.' Williams.

The Apollo 1 fire resulted in the cancellation of Apollos 2B and 3. Schirra, Eisele and Cunningham were assigned to Apollo 7 which would fly the intended Apollo 1 mission. Their back-ups were to be Stafford, Young and Cernan. Apollo 503, or Apollo 8, would be a test of all three Apollo modules in Earth orbit with McDivitt, Scott and Schweickart being backed up by Conrad, Gordon and, replacing Williams, Alan Bean.

Apollo 504, or Apollo 9, would be a repeat of Apollo 8 but in deep Earth orbit. This would be crewed by Borman, Collins and Anders. This is when Neil Armstrong made his appearance as back-up commander, with James Lovell and Buzz Aldrin. In 1968, it became clear that the Apollo 8 lunar module would not be ready in time and NASA decided to fly the mission into deep space or even around the Moon without it. McDivitt was offered the flight. His refusal resulted in Neil Armstrong, and not Pete Conrad, becoming the first man on the Moon. Borman's crew was assigned to Apollo 8 and McDivitt's to Apollo 9. The back-ups switched too, making Armstrong, Lovell and Aldrin eligible for Apollo 11. To complicate things, Collins had to drop himself from Apollo 8 to be replaced by Lovell, while Aldrin moved up to back-up command module pilot William Anders and Fred Haise came in as back-up lunar module pilot. By the time the Apollo 11 crew was chosen, Collins had been drafted into the command module pilot role and Aldrin reverted to the lunar module pilot position.

At the other end of the programme, there were to have been an Apollo 18, 19 and 20. Potential targets had been chosen, including some spectacular sites such as next to the crater Censorinus, the rim of the rayed crater

Joe Engle (left), in his days as an X-15 pilot. He was to have been lunar module pilot of Apollo 17 but lost his seat – and place in history – to Jack Schmitt who would have flown Apollo 18 had not budget cuts axed the mission.

Tycho in the southern hemisphere, the volcanic domes of the Marius Hills, Schroter's Valley, the 200 feet deep Hyginus Rille and inside the crater Copernicus.

Apollo 18 was to be commanded by Richard Gordon, former Apollo 12 command module pilot. His command module pilot was Vance Brand and the unique lunar module commander was Harrison Schmitt, a professional geologist as well as a bona fide NASA astronaut. Apollo 19's commander was Fred Haise, vet of the Apollo 13 drama. William

The Apollo 1 crew which perished in the fire on 27 January 1967. From left to right: Edward White, Gus Grissom and Roger Chaffee.

Apollo 2 was to have been commanded by Deke Slayton, seen here in space – at last, in 1975 – but he never got cleared to fly it and it was cancelled anyway.

Pogue was the command module pilot, and the sixteenth man on the Moon was to be Gerald Carr. Then, bringing up the rear, were Apollo 20's Conrad, Weitz and Lousma.

When the budget cuts slashed these flights, Schmitt ousted Apollo 17's Joe Engle from the coveted lunar module pilot's seat and Conrad, Pogue, Carr and Lousma hijacked the Skylab programme, ousting Cunningham, Schweickart, Bruce McCandless and Don Lind from their seats on America's first Space Station.

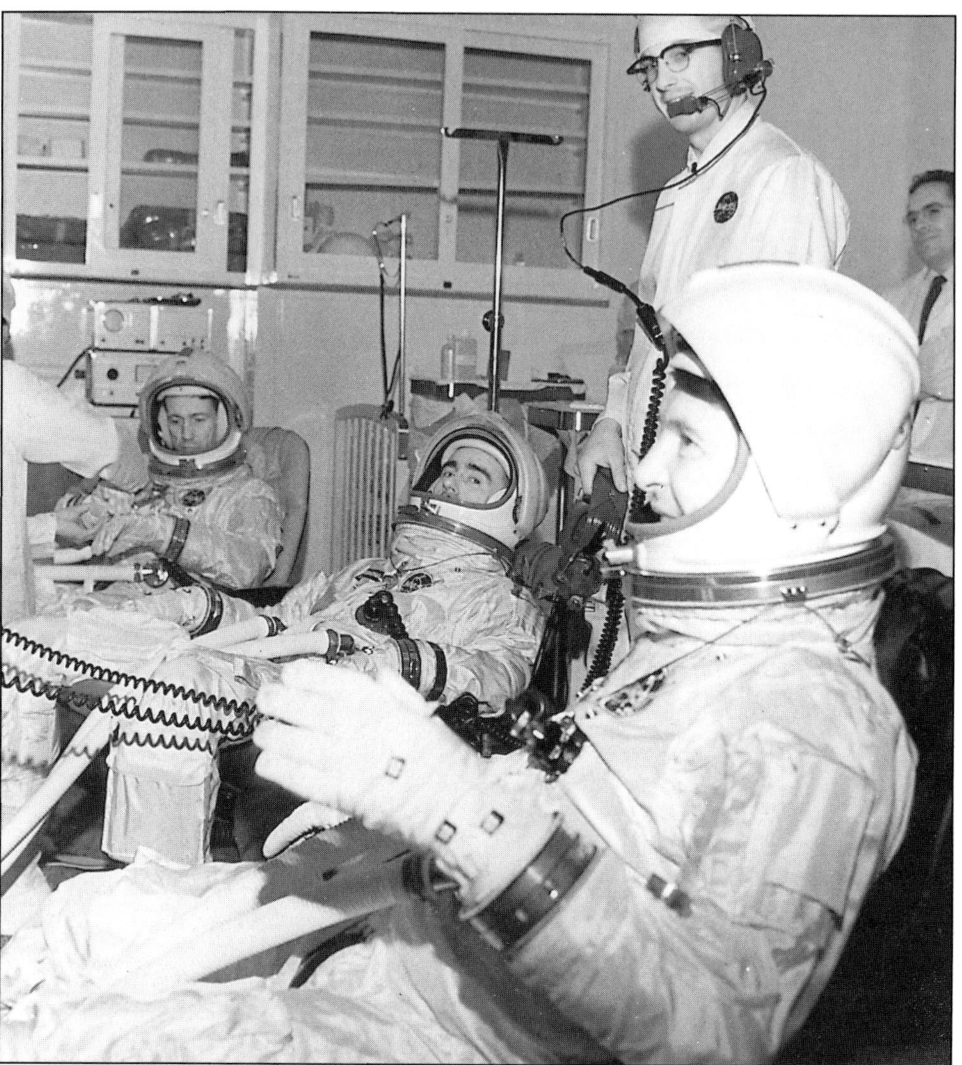

Wally Schirra, foreground, stood in for Slayton for Apollo 2 expecting to stand down for him. Also in the crew were Donn Eisele (left), and Walter Cunningham. Apollo 2 was cancelled and this crew eventually backed up Apollo 1, flying Apollo 7 in 1968.

Clifton C. Williams, who would have been the fourth man on the Moon, was killed in an air crash in 1967.

The Command Module Pilots

Mike Collins

In contrast to his companions, the affable, easy going Mike Collins provided some of the lighter moments during the Apollo 11 mission. He was offered the chance to serve as a back-up commander, possibly taking Apollo 17 to the Moon, but decided to leave NASA in 1970. He worked as assistant secretary of state for public affairs, director of the National Air and Space Museum in Washington DC, secretary of the Smithsonian Institution and vice president of the Vought Corporation, before establishing his own company. He wrote what has been considered the best astronaut biography, *Carrying the Fire*, and has followed this up with an authoritative history of the US manned space programme.

Mike Collins (centre), of Apollo 11.

Collins was born in Rome, Italy on 31 October 1930. After flying for the US Air

Force, Collins became a test pilot at Edwards Air Force Base and joined the third NASA astronaut group in October 1963. He was back-up Gemini 7 pilot and flew the Gemini 10 mission in July 1966 during which an Agena 10 restart took him and his colleague John Young to a then record altitude of 475 miles. Collins also became the first man to make bodily contact with another object in space during a curtailed spacewalk to retrieve samples from Agena 8. Collins was assigned as back-up pilot on Apollo 2A, which meant he was being groomed as a lunar module pilot. Then he was reassigned to Apollo 3 as command module pilot. This then became Apollo 9, then 8 but before he could fly to the Moon, he entered hospital for major spinal surgery. When he became available again, he was immediately drafted into Apollo 11. Collins is married to Patricia and has three children.

Dick Gordon

Thanks to budget cuts, US Navy Captain Richard Francis Gordon lost the chance of becoming the thirteenth man to set foot on the Moon as commander of Apollo 18. Gordon was born in Seattle, Washington on 5 October 1929. He was pilot on Project Bullet, which set a trans-American speed record and thus won the Bendix Trophy. From duties as a test pilot he joined the third NASA group in October 1963 and his first assignment was as back-up pilot to Gemini 8. He flew with his good friend Charles Conrad on Gemini 11 in September 1966. During this mission, thanks to an Agena restart, Gordon and Conrad reached a record earth orbital altitude of 850 miles over Australia. Gordon also performed a curtailed space walk. He was then assigned as back-up command module pilot to Apollo 3 and was reassigned in the same post to Apollo 8 which then became Apollo 9, thus making him elegible for Apollo 12. Gordon left NASA in 1971 to become vice president of the New Orleans Saints football team and is now president of Astro Sciences Corporation. He had six children from his first marriage, one of whom died. His second wife is Linda.

Dick Gordon (centre), with the Apollo 12 crew.

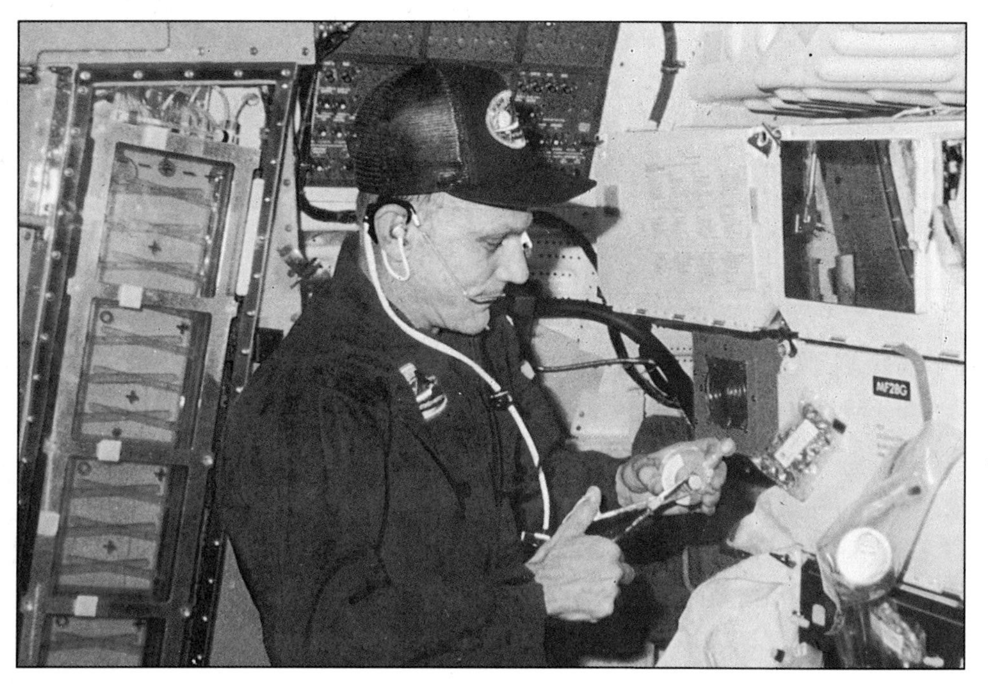

Ken Mattingly

The prematurely balding and gaunt looking Thomas Kenneth Mattingly II will be remembered for the German Measles he did not catch, as much as for the three spaceflights he made. Born in Chicago, Illinois on 17 March 1936, Mattingly flew for the US Navy and was a student at the aerospace research pilots school at Edwards when selected to join the 1966 Group 5 astronaut intake. He served on three support crews in Apollo before being assigned as Apollo 13 command module pilot. Two days before lift off in April 1970, he was dropped from the mission when it was discovered that Charlie Duke has

Ken Mattingly aboard the Shuttle in 1982.

exposed him to German measles and there was a fear that he would develop the mild illness in space. He never did catch it and is maybe glad he did not because as Apollo 16 command module pilot he was associated with the successfull landing – not before scares with the CM engine – and had the bonus of a spacewalk on the way home.

Moving on to the Shuttle programme, Mattingly was regarded as one of the prime candidates for an early flight. He served as back-up commander of STS 2 and 3, for which he and Hank Hartsfield were the last formal back-ups in NASA history. He flew two military Shuttle flights, STS 4 in June 1982 and STS 51C in January 1965. He left Nasa soon after to become commander of the US Navy Electronics Systems Command. He has been divorced twice and has one son.

Al Worden in 1984.

Stuart Roosa

Stuart Allen 'Smoky' Roosa, a ginger-haired US Air Force Colonel, was the most unheralded of the lunar landing command module pilots. He was overshadowed by the publicity surrounding his commander, Alan Shepard, and was not as busy in lunar orbit as later CMPs, who also had the chance to walk in space on the way home. His starring role was the multi-attempted docking with the lunar module *Antares* in the S4B stage during trans-lunar injection. Roosa was born in Durango, Colorado on 16 August 1933 and was an accomplished test pilot, qualifiying as an aerospace research pilot at Edwards Air Force Base before his selection to the fifth NASA astronaut group in April 1966. He was named for Apollo 14 without having served on a back-up crew. After Apollo 14 in January/February 1971, Roosa then did his back-up stints for Apollo 16 and 17 and left NASA in 1976. He worked in Athens, Greece for a short while before becoming president of Jet Industries, then – perhaps with the help of his Apollo 14 commander – became president and owner of a Gulf Coors beer distributorship in Mississippi. He is married to Joan and has four children.

Stu Roosa at Coors Beer.

Al Worden

Alfred Merrill Worden became the first man to make a trans-Earth EVA during the Apollo 15 mission in July/August 1971. In lunar orbit he said he "felt like a UFO driver, spying on someone's back yard. You get the feeling that you shouldn't be there". Worden was born on 7 February 1932 in Jackson, Michigan and attended the Empire Test Pilot School at Farnborough in England before eventually becoming a test pilot instructor at Edwards Air Force Base. Worden joined the fifth NASA astronaut group in 1966 and served on the back-up crew of Apollo 12. He was dropped from the Apollo 17 back-up crew, together with his Apollo 15 colleagues for disciplinary reasons and left the astronaut group in 1972 and NASA in 1975. He works with fellow astronaut Jim Irwin in the High Flight Foundation and is now president of his own company in Florida. The father of two, he divorced, then remarried, fathering one more child before divorcing again.

Ron Evans

Born in St Francis, Kansas on 10 November, 1933, Ronald Ellwin Evans was a US Navy Vietnam veteran, flying F-8s from the *USS Ticonderoga*. He joined the NASA group 5 team in 1966, serving on three Apollo support teams before being assigned as back-up for Apollo 14. During his December 1972 flight, Evans performed the now customary trans-Earth EVA. He was the back-up command module pilot of the Apollo-Soyuz mission in 1975 and resigned two years later. He works for Sperry Flight Systems. He and his wife Janet have two children.

Ron Evans (centre), in raft after Apollo 17.

The Apollo Pathfinders

Gus Grissom

Many people thought that, had he lived, little Gus Grissom would have been the first man on the Moon. They were probably right. Living up to his 'gruff Gus' image, the astronaut's final public words were, "How do you expect us to get to the Moon if you can't talk between two buildings?". These were spoken at about 6.30 pm on 27 January 1967 from inside the command module of Apollo 1, sitting atop un unfuelled Saturn 1B at Pad 34, Cape Kennedy, as it was then called. The simulated countdown had been put on hold yet again, this time by a communications problem. A minute or so later he was choked to death with two colleagues in the infamous Apollo fire. His body lies at Arlington military cemetary in Washington, where it was buried with full honours in the presence of President Johnson.

Virgil Ivan Grissom was born in Mitchell, Indiana on 23 April 1926 and studied mechanical engineering before becoming a US Air Force pilot, flying 100 combat missions in F86 Sabres during the Korean War. Grissom then received a degree in aeronautical engineering and in 1959 ended up one of the famed Mercury Seven as a result of the vagaries of a selection system that failed to screen all available test pilots, because so many volunteered. He earned his public 'gruff Gus' image with his performance at a press conference on 9 April 1959.

Gus Grissom in his Mercury days.

In January 1961, he was named as one of three astronauts chosen to compete for the first Mercury mission, a suborbital affair on a Redstone IRBM, or to fly one of the two later missions. He got the second slot and on 21 July 1961 became the third man in space, reaching a height of 118.5 miles in his spacecraft *Liberty Bell 7*. Afterwards the great mystery occurred. The spacecraft's hatch blew open, seawater rushed in and Grissom rushed out, nearly drowning in the ensuing drama, while a helicopter tried vainly to stop *Liberty Bell* from sinking. Grissom always maintained that he "was just lying there when it blew". In the film *The Right Stuff*, based on Tom Wolfe's portrait of the Mercury Seven, Grissom is depicted, rather unkindly, in a state of panic after the flight and in a hurry to get out of the capsule. In any event, the astronaut, who admitted that he was scared at blast-off, and who lost his spaceship became the first commander of the Gemini programme, after the first choice, Alan Shepard, was declared unfit. He was named for the Gemini 3 mission on 13 April 1964 and took to the skies at Pad 19 Cape Kennedy on 25 March the following year, atop a Titan 2 ICBM. His three-orbit flight, with John Young, was the first to involve in-orbit manoeuvring. He was the first man to make two spaceflights. Grissom served as back-up commander for Gemini 6 and in March 1966 was named to command Apollo 1. He was married to Betty, had two boys and was a Lieutenant Colonel at the time of his death.

Ed White

In contrast to Grissom, Apollo 1 senior pilot, Lt Col Edward Higgins White II, USAF, was a charismatic space hero, a West Point graduate who became the first American to walk in space, featuring in some of the most spectacular photographs ever taken in space, many of which are still used today. White failed to make the Mercury team but joined the second class of NASA astronauts in September 1962.

White, the son of an Air Force general, was born in San Antonio, Texas on 14 November 1930, and followed his father's footsteps, gaining degrees in science and engineering, joining the USAF and becoming a test pilot. Before becoming an astronaut, he flew some Boeing 707 zero-g parabolas for astronaut training. White was assigned as pilot of Gemini 4 and was probably going to perform an EVA, standing on his seat, with the spacecraft hatch open. Soviet cosmonaut Alexei Leonov's first EVA in March 1965, changed this plan. White trained for a full EVA in which he would use a hand-held manoeuvring unit. He was launched, with James McDivitt, on 3 June 1965. During this four-day flight, White walked in space for 22 minutes, capturing the headlines all over the world. He was then assigned as back-up commander of the long-duration Gemini 7 mission before moving on to Apollo 1. White, who was married to Pat and had two children, was buried at West Point.

Ed White on Gemini 4 launch day.

Roger Chaffee

The third victim of the infamous fire, baby-faced Roger Chaffee was preparing for his first spaceflight. Beneath his youthful image, was a highly ambitious and qualified astronaut, the first to be chosen for a mission from the third group of NASA spacemen selected in October 1963, but only after first-choice Donn Eisele was injured. Lt Cdr Roger Bruce Chaffee, USN, was born in Grand Rapids, Michigan on 15 February, 1935. He gained degrees in aeronautical engineering and in engineering. Chaffee flew reconnaisance mission over Cuba during the world crisis in October 1962. He was married to Martha and had two children. Had he flown the Apollo 1 mission, he would still today have been the youngest American in space. Chaffee is buried at Arlington Cemetary, Washington.

Roger Chaffee on selection in 1963.

Wally Schirra

It may have been galling for the fun-loving, joking, enthusiastic Walter Marty Schirra Jr to fly the most anonymous of the six manned Mercury missions, a modest six-orbit affair on 3 October 1962 but he certainly made up for it on his next two flights. The commander of the Gemini 6 mission which performed the first rendezvous in space, with Gemini 7, after launch on 15 December 1965, Schirra played jingle bells and reported seeing a UFO called 'Santa Claus'. His good nature shone through again when he commanded the Apollo 7 Earth orbital shakedown flight, starting on 11 October 1968 but soon evaporated after he caught a cold and issued verbal tirades to the ground. He insisted on becoming the on-board flight director and his heavy-handed attitude lost him some of the great respect he had earned in his career.

Offered a moon flight and the chance to become the first astronaut admiral, Schirra resigned his NASA and US Navy commissions. Schirra's personality made him a hit during TV specials on the Apollo moon flights and he was so overcome with emotion after Armstrong's touchdown that he shed a few tears on air with Walter Cronkite. Schirra was born in Hackensack, New Jersey on 12 March 1923 and flew 90 combat missions for the US Navy, downing two MiGs and earning a DFC and air medals. He became a test pilot before joining the Mercury team, pulling off many legendary 'Gotcha' jokes on his colleagues and associates during training. Before flying Apollo 7, he had been assigned to the second manned Apollo flight, Apollo 2A, which was then cancelled and he was reassigned as back-up commander of Apollo 1. After leaving NASA he entered private business, eventually forming his own consultancy. He is married to his 'original wife', as he calls

Wally Schirra, pilot of Mercury capsule Sigma 7.

Donn Eisele, in 1966, died in 1988.

Walter Cunningham in 1984. (Walter Cunningham)

Jo (after the 'Original 7' Mercury astronauts), and has two children. He is the first man to make three spaceflights and only astronaut to have flown Mercury, Gemini and Apollo missions.

Donn Eisele

The second astronaut to die of natural causes, Donn Eisele died alone, of a heart attack, aged 58, in a hotel room in Tokyo on 1 December 1987, either before or after a jog. Fellow astronaut Alan Shepard was also visiting Tokyo with Eisele and was on hand to make arrangements. Eisele was an executive of the Oppenheimer and Co. investment firm. Though fortunate in missing the Apollo 1 disaster through injuring his shoulder in weightless training, in a sense, Donn Eisele had the misfortune to be chosen as senior pilot for Apollo 7. The highly capable astronaut was totally overshadowed by Schirra during the mission and, through astronaut politics, never flew again. Eisele, who became the first active astronaut to divorce, shortly after Apollo 7, was assigned to back-up command module pilot of Apollo 10 in 1969. the commander of this back-crew was Gordon Cooper, who was not playing the game had he really wanted a later moonflight. As a result Cooper was not assigned Apollo 13 and Eisele missed out, which for this fated astronaut may have been fortunate. Before Apollo 7, he had originally been assigned to the cancelled Apollo 2A then as Apollo 1 back-up. He left the astronaut corps, was reassigned in NASA then left in 1972, to become a director of the Peace Corps in Thailand. Col Donn Fulton Eisele, USAF, Rtd, was born on 23 June 1930 in Columbus, Ohio and was a former aerospace research pilot, joining the third NASA astronaut group in 1963. He left six children, four from his first marriage and two from his second marriage to Susan.

Walt Cunningham

Like Eisele, Cunningham, who put in a creditable performance on Apollo 7 was overshadowed. He was born in Creston, Iowa on 16 March 1932 and flew as a US Navy pilot before receiving a doctorate in physics and working for the Rand Corp. He was somewhat out of place in the third astronaut group chosen in October 1963. He was first assigned as pilot of the second Apollo flight, designated Apollo 2A, but when this was cancelled was reassigned to back-up the Apollo 1 crew, before taking the mission after the Apollo fire. Politics played a part in his astronaut career, too. Cunningham moved straight from Apollo 7 to the Apollo Applications Programme, later to become Skylab. He was seen widely as the commander of the first manned occupation until some later Apollo moonflights were cancelled and Pete Conrad lost the chance of becoming the only man to make two moon landings. Conrad took over the Skylab office and Cunningham lost out, being offered the back-up's job on the third and last mission. One of the greatest examples of the victimization of astronaut politics, it wasn't surprising that Cunningham resigned two years before Skylab was launched. Married to Lo, he has two children and runs a private banking and investment firm in Houston.

Frank Borman

If there is one astronaut who many felt deserved a landing on the Moon, it was Frank Borman. He probably would have been assigned to such a flight had he not decided to resign in 1970, having reached the pinnacle of his career, as one of the first men to fly to and around the Moon, on the epic journey of Apollo 8 in December 1968. The quintessence of an astronaut, the no-nonsense, quick-thinking and disciplined US Air Force Colonel was born in Gary, Indiana on 14 March 1928. He graduated to become an instructor at the USAF Aerospace Pilots' school at Edwards Air Force Base. Borman joined the second astronaut group in 1962.

Frank Borman with Eastern Airlines in 1984.

He was first choice to pilot Gemini 3 with Alan Shepard but lost the chance when Shepard became ill. Instead he served as back-up commander of Gemini 4 and commander of Gemini 7, a marathon fourteen-day record-breaking mission in December 1965. Borman then moved straight into Apollo being named as back-up commander on the cancelled Apollo 2A mission. He was then assigned prime commander's seat on Apollo 3 which was to test a lunar module launched separately on a Saturn 1B. This mission later became an all-up job on the mighty Saturn 5 during which Borman would test the lunar module in deep Earth orbit. Delays to the development of this lunar module meant that Apollo 9, as it was called, would fly as Apollo 8, without it. Apollo 8 then became a Moon orbit mission because of fears that the Soviets were about to send one cosmonaut on a lunar looping mission in late 1968.

Borman, who is married to Susan and has two children, joined Eastern Airlines in 1970, eventually becoming its chairman, president and chief executive officer. After Eastern was taken over in 1986, Borman retired. In 1988, he reappeared on the scene, being appointed chairman and chief executive of the Patlex Corporation, a company holding several valuable laser patents.

Jim Lovell

Jim Lovell had been sidetracked in the Gemini programme as back-up commander of Gemini 10 as some of his flight-experienced colleagues moved on to Apollo. The Gemini programme ended at 12 and Lovell would have commanded 13. It took the deaths of two astronauts, however, the injury of one and astronaut politics to make Lovell the first to fly two Apollos. Because the fatalities were the prime Gemini 9 crew, Lovell and his colleague Buzz Aldrin were moved to back-up this mission, thereby being eligible for Gemini 12 in November 1966, his second mission, following his Gemini 7 marathon in December 1965. Following this, Lovell went on to the back-up crew of what was originally Apollo 9 but which became Apollo 8.

Injury to Mike Collins gave Lovell the command module pilot's seat of the famed Apollo 8 mission to the Moon in December 1968 and the back-up commander role for Apollo 11. Normally, he would have been eligible to command Apollo 14 but because Alan Shepard needed more training time, Lovell got Apollo 13 and almost met a lingering death in space. The first man to make four space missions, Capt James Arthur Lovell Jr, USN, was born in Cleveland, Ohio on 25 March 1928 and joined the second group of NASA astronauts in September 1962, serving first as back-pilot for Gemini 4. He is married to Marylin, has four children and is group vice president of the business communications group of the Centel Corporation.

Bill Anders

As lunar module pilot of Apollo 9, Anders was to fly into deep Earth orbit in the flimsy spacecraft, to simulate a Moon landing. However, he was switched to Apollo 8 which had no lunar module and although he was one of the first men to fly the Saturn 5 and to orbit the Moon, Anders was disappointed to fly the LM. William Alison Anders was born in Hong Kong on 17 October 1933 and had a degree in nuclear engineering as well as being

James Lovell, right.

Bill Anders after Apollo 8. (Eileen McClure from Bill Anders)

an Air Force pilot. He was among the third group of astronauts selected just after his thirtieth birthday. He served as back-pilot on Gemini 11 before moving to Apollo 3, which was later redesignated Apollo 9 and eventually became Apollo 8, which flew in December 1968. This was Anders' only spaceflight, although he did act as back-up command module pilot of Apollo 11 and would have flown as the CMP on Apollo 13 had he not resigned in 1969. He was chairman of the US Nuclear Regulatory Commission and the US Ambassador to Norway before eventually becoming vice president for operations for Textron Inc. He is married to Valeri and has six children. He is also a Major General in the US Air Force Reserve.

Jim McDivitt

Former plumber, Brigadier General James Alton McDivitt USAF, Rtd, is executive vice president of the Defence Electronics Operations Group of Rockwell International. At the time of his resignation from NASA, McDivitt was director of the Apollo programme, having been manager of the lunar landing operations group at the time of the Apollo 11 landing. This landing was made possible by McDivitt's piloting of the first lunar module during the Earth orbital test flight of Apollo 9 in March 1969. Due to schedule changes, McDivitt could have flown the Apollo 8 lunar mission but was determined to keep his original assignment. Prior to Apollo 9, McDivitt had been assigned to a cancelled mission called Apollo 2B having moved from the back-up crew of Apollo 1 after a few months. He moved to Apollo after commanding Gemini 4 to which he was assigned two years after becoming an astronaut with the second group in 1962.

Jim McDivitt, Rockwell executive in 1984. (Rockwell)

McDivitt was born in Chicago, Illinois on 10 June 1929 and flew 145 combat missions in the Korean War earning five air medals and a DFC. He was a test pilot and aerospace pilot at Edwards Air Force Base and was almost chosen for the X-15 rocket plane programme. McDivitt is the divorced father of four children.

Rusty Schweickart

Former MIT scientist Russell Louis Schweickart seemed out of place in the test piloting role of lunar module pilot of Apollo 9 in March 1969. He was, however, a former USAF pilot. He was born in Neptune, New Jersey on 25 October 1935 and joined the third astronaut group in 1963. He missed out on Gemini, becoming first the original Apollo 1 back-up pilot and was subsequently assigned as lunar module pilot of Apollo 2B, which then became Apollo 8, then 9, which flew in March 1969. Space sickness deprived him of the chance of trying out the lunar spacesuit and portable life support system back-pack during a spacewalk from the lunar module to the docked command module in Earth orbit. Instead, he remained standing on the porch of the LM. By flying the LM, called *Spider*, independently in Earth orbit, with his commander McDivitt, Schweickart became the first person to fly in space in a spacecraft that was not capable of flying back to earth. After Apollo 9, Schweickart was assigned to the Skylab programme and served as back-up commander of the first mission in 1973. He remained with NASA, involved in the Shuttle but out of the crew mainstream and left in 1979 to become chairman of the California Energy Commission. He is married to Clare and has five children.

Rusty Schweickart in 1984.

Fred Haise

In 1979, Shuttle flight 3 docks with the Skylab space station in a dramatic bid to stop it plunging to Earth. A rocket stage is attached and fires, placing Skylab into a higher and safe orbit. This was to have been the highlight of Fred Haise's astronaut career as commander of OFT 3. It never happened and Skylab plunged to Earth anyway. The Apollo 19 lunar module touches down on the Moon in 1973 and Haise walks on the Moon. That never happened because of budget cuts. Fred Wallace Haise was born in Biloxi, Mississippi on 14 November 1933 and flew for the US Navy, Air Force and Marine Corps before becoming a NASA test pilot. He joined the fifth astronaut group in April 1966. Haise supported the Apollo 1 mission and was back-up lunar module of Apollo 8. Normally he would have been assigned to Apollo 11 but as he hadn't flown a space mission he was replaced and assigned to back up Aldrin on the first landing mission and then continued in that position as prime lunar module pilot of Apollo 13 to play a role in that drama. He was back-up commander of Apollo 16 and would have commanded Apollo 19 had not the mission been cut by the budget axe. Haise was a leading Shuttle astronaut, flying in 1977 three *Enterprise* approach and landing test missions into Edwards Air Force Base.

With his 1979 Shuttle mission delayed and with no prospect of a flight for another three years, Haise resigned and joined Grumman Aerospace, where he is now a surprisingly portly president of the company's technical services division, with a key role in Shuttle ground operations. Haise is married to Mary and has four children.

Fred Haise.

Tom Stafford

Had his lunar module been capable of landing on the Moon, Tom Stafford could have become the first man on the Moon. Instead, he came to within nine miles of the surface during the Apollo 10 test flight in May 1969.

Tom Stafford.

A cartoon by Franklin depicted him hanging out of the lunar module, *Snoopy*, with his toe stretching for the surface, saying "It's just so that I can say I touched it!". He also became one of the few astronauts to make colourful conversation: "There's Censorinus," he said of a crater, "and it's bigger than shit!" In July 1975, his hand stretched for the hand of Soviet cosmonaut Alexei Leonov, during the first and so-far only Soviet-US space mission, ASTP. Lt General Thomas Patten Stafford, USAF Rtd, was born in Weatherford, Oklahoma on 17 September 1930 and joined the

second group of NASA astronauts from Edwards Air Force base where he was one of the leading test pilots. He worked for a while as pilot to Alan Shepard for Gemini 3 but when Shepard became ill, joined Wally Schirra on the back-up crew, later flying the Gemini 6 mission in December 1965 only after a heart-stopping launch pad abort. Staford was originally back-up command pilot of Gemini 9, a mission he flew after the death of the prime crew in an air crash. He was then assigned back-up senior pilot of Apollo 2A then back-up commander of Apollo 2B. Both these missions were soon cancelled and after the Apollo 1 fire Stafford became back-up commander of Apollo 7. He left NASA in 1975 and became commander of the USAF Edwards Air Force Base. He is now vice president of Gibraltar Exploration in his native Oklahoma where he lives with wife Faye. They have two children.

Jack Swigert

Reknowned bachelor and ladies' man, John Leonard 'Jack' Swigert was born in Denver, Colorado on 30 August 1931 and was a US air Force Korean War veteran. He became a test pilot for Pratt and Whitney and one of two civilian astronauts in the fifth NASA group chosen in April 1966. He supported four Apollo missions before being assigned as back-up command module pilot of Apollo 13 with the likelihood of flying the Apollo 16 mission to the Moon. Two days before Apollo 13 took off, the suspected impending illness of prime crewman Tom Mattingly, gave Swigert the flight seat before he had a chance to fill in his tax papers in April 1970. Two days later he was chief character in the famous saga of Apollo 13, uttering the first, casual words from the crippled command module, which have since become as famous as Armstrong's first words on the Moon: "Hey, Houston, we've had a problem". Swigert took leave-of-absence from NASA

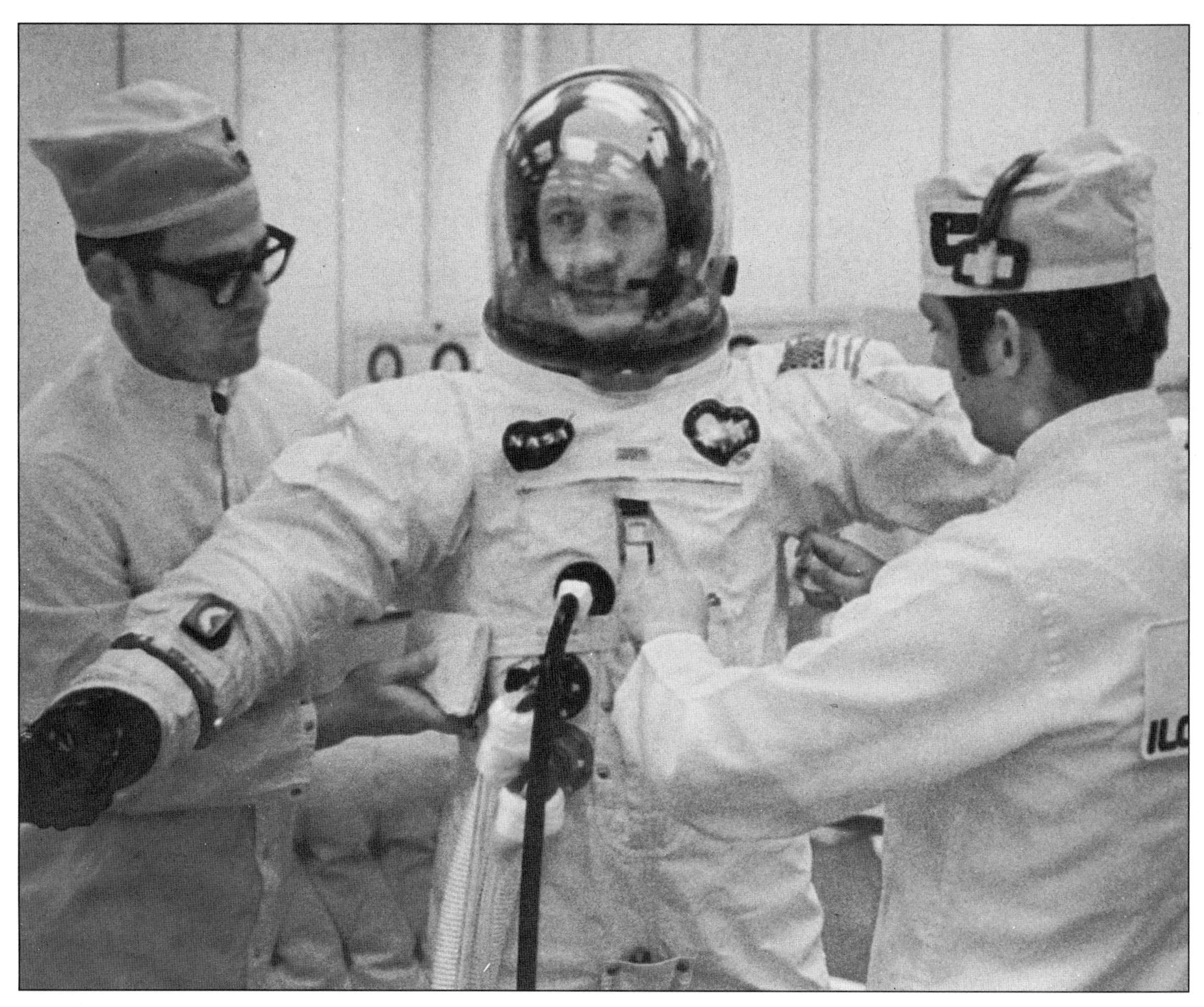

Jack Swigert.

shortly after arriving home to receive the NASA distinguished service medal. He could have rejoined to become Shuttle pilot but attempted to enter politics, discovering during his last campaign that he had bone cancer. Successfully elected to office and one week before he was to enter the House of Representatives he died, on 27 December 1982. The first US astronaut to die of natural causes, Swigert never married.

The Chosen Dozen

THE OLDEST MAN ON THE MOON WAS Alan Shepard, aged 47. The youngest, Charlie Duke, was 36. The twelve men who walked on the Moon were middle-aged. Whether such a thing really exists or is a figment of the individual's imagination, each of the twelve had the opportunity to experience a 'mid-life crisis'. The onset of 40 in a man is the source of constant mirth to his friends, and perhaps his enemies. It is a time when somehow you look back at what you've done – some of it good, some of it bad; some of it pleasant, some of it unpleasant. You realize that probably half your life has gone. You realize the difference between what you want to do and what you want to do that you can't. You know what you like doing and what you don't like doing. Maybe, there's less aggression about the way you do things. Altogther, you've got your act together and you know where you stand in the scheme of things.

For some people this can be a calming influence but it can make others feel at odds with the world. The Twelve included people of both types and the mid-life crisis that some experienced was exacerbated by the feeling of having achieved an ultimate goal, with nowhere to go but down. It was these that searched for an answer to life after the Moon. Some found the answer and some are still searching, maybe never to find it. And then, there are the others who sailed serenely through it all. Armstrong knew exactly what he wanted to do and did not let the Moon change it for him. Aldrin was profoundly effected by the Moon. He'd reached the goal and he rummaged through life trying to find solitude and serenity. Conrad breezed through, the same guy now that he was before he flew to the Moon. In some senses Bean experienced a mid-life crisis but found the answer on his doorstep – painting – something he had been doing and liking even before the Moon. Shepard. Well, as he says, "I was a real son of a bitch before I went to the Moon; now, I'm just a son of a bitch". He did A-OK did Al, whether people liked it or not. Mitchell experienced pangs of moon-life crisis, never really finding the answers. Scott is another who apparently found life after the Moon a breeze. Irwin, though, was reawakened by the truth of the Lord who he met, again, on the Moon, in a most profound, intense and meaningful way. He also suffered ill-health, almost losing his life with heart attacks. Young will always be an astronaut whether he flies to the Moon, rides the Shuttle or drives a car. They'll have to pull off his space boots because he'll not do it himself. Duke is another who found the Moon a difficult act to follow. He searched and found God. Cernan made no excuses about going to the Moon and made the most of it. Schmitt, too, used the Moon to great effect.

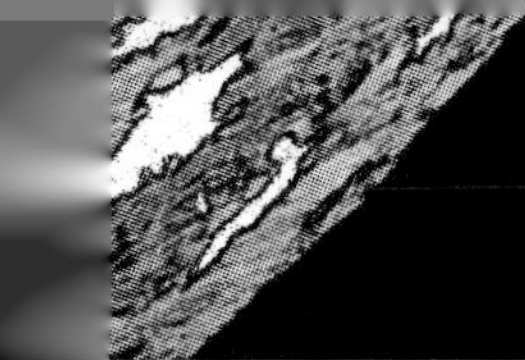

Earth Rise to Full Earth

Twelve men have walked on the Moon, but it is, probably, these images of Mother Earth that had the most profound and lasting effect on each of them.

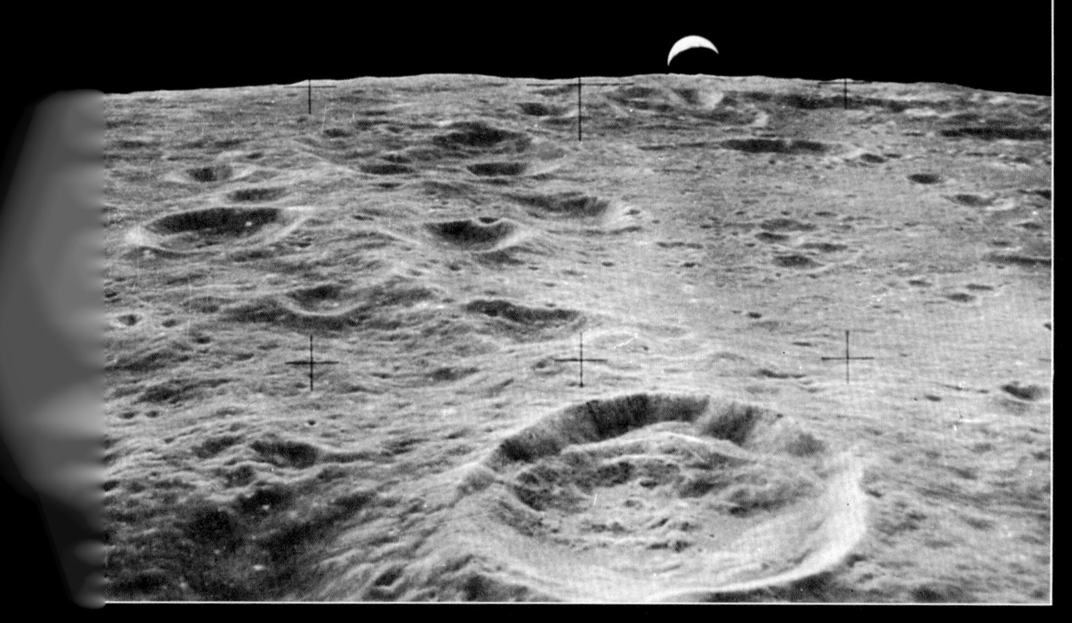

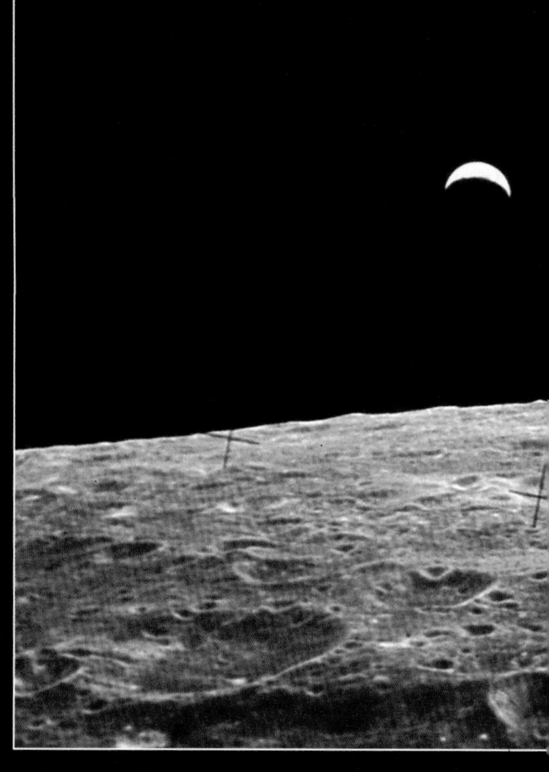

But the truth is, that each of them was the same personality before he went to the Moon and probably would have done pretty much the same as they did after the flight. Armstrong would have become a professor and a reclusive businessman. Aldrin was a shy and introverted person. Conrad would have gone into the aviation business. Bean would probably find painting a major activity in his life. Shepard had things mapped out pretty well before he went. Scott was heading to a business career. Irwin was a Christian before he went. Young was going to keep his spaceboots on for ever. Duke would have done something to earn a lot of money for his family, and his Christian wife would have been an influence. Cernan was heading for the Cernan Corporation with or without the Moon. Schmitt was certainly heading for the US Senate.

If some were doing before the moon-flight what they ended up doing after, they would never have flown there in the first place. For example, if Jim Irwin was an astronaut evangelist or Ed Mitchell an active and open ESP nut, NASA wouldn't have flown them. They might have done in the pre-*Challenger* Shuttle era but certainly not in the climate of Apollo. The difference, perhaps, is how the Moon focused attention on these ambitions, faults, aspirations and how it helped some to find themselves more easily and with more clarity of purpose and with more fulfilment than they imagined they could. In any event, it could have been twelve other astronauts; it just happened to be these. They were also figureheads of a grand American enterprise involving, directly and indirectly, milllions of countrymen who would not have appreciate any of them taking all the credit and glory.

It is no coincidence, however, that the twelve came back from the Moon with an image forever etched in their minds of the small planet Earth, ravaged by man-made pollution, war and inequality, alone in the vastness of space; an image brought to us through the photographs they brought back. Each experienced a realization that someone out there is watching over the Earth; The Universe. Photographs may impart some of that feeling but it is the Twelve who experienced the ultimate realization. It is perhaps, therefore, also no coincidence that, like Christ's Apostles, there were Twelve of them. That, perhaps, is the message of their journeys – to the Moon.

Appendices

A. Flotsam and Jetsam

ELEVEN APOLLO COMMAND MODULES, THE descent stages of six lunar modules, the ascent stages of two lunar modules and three lunar rovers have survived the Moon programme. The debris of five crashed ascent stages litter the Moon. The command modules are on display at various exhibitions. The other hardware is a little less accessible. One day, perhaps, Eagle's descent stage, a flag from the Apollo 11 mission or even a lunar rover may be recovered and brought back to Earth.

The Command Modules

Apollo 1: In sealed storage at Langley Air Force Base, Virginia, USA.
Apollo 7: National Museum of Science and Technology, Ottowa, Canada.
Apollo 8: Chicago Museum of Science and Technology, Chicago, Illinois, USA.

The following command modules had names.
Apollo 9, *Gumdrop*: Michigan Space Centre, Jackson Community College, Jackson, Michigan, USA.
Apollo 10, *Charlie Brown*: Space Exploration exhibition, Science Museum, London, England.
Apollo 11, *Columbia*: National Air and Space Museum, Washington DC, USA.
Apollo 12, *Yankee Clipper*: NASA Langley Research Centre, Langley, Virginia, USA.
Apollo 13, *Odyssey*: NASA Kennedy Space Centre, Cape Canaveral, Florida, USA.
Apollo 14, *Kitty Hawk*: Rockwell International, Downey, California, USA.
Apollo 15, *Endeavour*: US Air Force Museum, Wright Patterson Air Force Base, Dayton, Ohio, USA.
Apollo 16, *Caspar*: Alabama Space and Rocket Centre, Huntsville, Alabama, USA.
Apollo 17, *America*: NASA Johnson Space Centre, Houston, Texas, USA.

Lunar Module Descent Stages

Apollo 10, *Snoopy*: In lunar orbit.
Apollo 11, *Eagle*: Tranquility Base, 0.41 deg N 23.36 deg E.
Apollo 12, *Intrepid*: Ocean of Storms, 3.11 deg S 23.23 deg W.
Apollo 14, *Antares*: Fra Mauro, 3.40 deg S, 17.28 deg W.
Apollo 15, *Falcon*: Hadley Base, 26.5 deg N, 3.40 deg E.
Apollo 16, *Orion*: Descartes, 8.59 deg S, 15.30 deg E.
Apollo 17, *Challenger*: Taurus Littrow, 20.10 deg N, 30.45 deg E

Lunar Module Ascent Stages

Apollo 10, *Snoopy*: In solar orbit.
Apollo 11, *Eagle*: In lunar orbit.
Apollo 12, *Intrepid*: Crashed on Moon, 20 November 1969.

Apollo 14, *Antares*: Crashed on Moon, 7 February 1971.
Apollo 15, *Falcon*: Crashed on Moon, 3 August 1971.
Apollo 16, *Orion*: Crashed on Moon, 29 May 1972.
Apollo 17, *Challenger*: Crashed on Moon, 15 December 1972.

Lunar Roving Vehicles
Apollo LRV 15: Parked at Hadley Base, 26.5 deg N, 3.40 deg E.
Apollo LRV 16: Parked at Descartes, 8.59 deg S 15.30 deg E.
Apollo LRV 17: Parked at Taurus Littrow, 20.10 deg N, 30.45 deg E.

The Apollo 17 command module America.

The Apollo 17 lunar rover vehicle parked at Taurus Littrow.

Above: The ascent stage of Eagle, *Apollo 11's lunar module.*

B. Reference Sources

Chapter 2: The Media Dream

1 Daily Mail
2 Daily Telegraph
3 Daily Herald
4 Daily Mail
5 Evening News
6 Daily Herald
7 Daily Telegraph
8 Daily Telegraph
9 Daily Express
10 Daily Mirror
11 Daily Mirror
12 Evening News
13 Daily Mirror
14 Daily Sketch
15 Sunday Express
16 Daily Telegraph
17 Evening Standard
18 Daily Telegraph
19 Evening Standard
20 Sun
21 Daily Mirror
22 Daily Mirror
23 Daily Express
24 Sunday Telegraph
25 Daily Express
26 Daily Express
27 Daily Mail
28 Daily Mirror
29 Daily Express
30 Daily Mail
31 Sun
32 Evening News
33 Daily Mail
34 Daily Express
35 Daily Telegraph
36 Evening News
37 Daily Express
38 Daily Mail
39 Daily Mail
40 Daily Express
41 Daily Express
42 Daily Telegraph
43 Daily Mail
44 Evening Standard
45 Sunday Mirror
46 Evening News
47 Daily Telegraph
48 Daily Mail
49 Daily Mail
50 Evening News
51 Evening Standard
52 Daily Express
53 Evening Standard
54 Evening Standard
55 Daily Telegraph
56 The Times
57 The Sun
58 Daily Mirror
59 Sunday Mirror
60 Daily Mail
61 Evening News
62 News of the World
63 Evening Standard
64 Evening Standard
65 Evening News
66 Daily Express
67 Daily Mail
68 Evening News
69 Daily Mirror
70 Financial Times

71 Evening News
72 Evening Standard
73 Daily Express
74 Daily Sketch
75 Daily Telegraph
76 Evening Standard
77 The Guardian
78 Daily Express
79 Daily Mail
80 Daily Mirror
81 Daily Express
82 Daily Sketch
83 The Times
84 Daily Mail

Chapter 3: The First Steps at Tranquillity

1 Daily Mail
2 Daily Mail
3 Daily Sketch
4 Daily Express
5 Daily Mirror
6 The People
7 Sunday Mirror
8 Sunday Mirror
9 Sunday Times
10 Times
11 Daily Sketch
12 Daily Telegraph
13 Sun
14 Daily Mirror
15 Evening Standard
16 Evening News
17 Daily Mirror

Chapter 4: Thunderstruck en route to the Ocean of Storms

1 Daily Express
2 Daily Mail
3 Sun
4 Times
5 Daily Sketch
6 Daily Express
7 Daily Telegraph
8 Daily Mail
9 Daily Express
10 Evening Standard
11 Evening Standard
12 Daily Sketch
13 Sun
14 Sun
15 Daily Telegraph
16 Evening News
17 Daily Mail

Chapter 5: The Main B Bus Undervolt

1 Evening News
2 Daily Mail
3 Evening Standard
4 Daily Mail
5 Daily Mail
6 Daily Mail

Chapter 6: Rookies Head for Fra Mauro

1 Daily Express
2 Evening News
3 Daily Express
4 The Sun
5 Evening News
6 Daily Express
7 Daily Mail
8 Daily Mirror
9 Daily Express
10 Daily Mail
11 Evening Standard
12 Evening Standard
13 Daily Mail

Chapter 7: The Rovers of Hadley Base

1 Evening News
2 Evening Standard
3 Daily Mail
4 Daily Mail
5 Daily Mirror
6 Daily Telegraph
7 Sunday Express
8 The Sun
9 Daily Mail
10 Daily Mail
11 Daily Express
12 Sunday Times
13 Sunday Express
14 Daily Telegraph
15 Daily Mail
16 The Guardian
17 Daily Express
18 The Guardian
19 Sunday Mirror

Chapter 8: The Jokers of Descartes

1 Financial Times
2 Daily Mail
3 The Guardian
4 The Times
5 The Times
6 Sunday Telegraph
7 The Times
8 Evening News
9 The Guardian
10 Daily Mirror
11 The Sun
12 Daily Mail
13 Daily Express
14 Evening News
15 Daily Express
16 Daily Mirror
17 Daily Mail

Chapter 9: The End at Taurus Littrow

1 Sunday Times
2 Sunday Times
3 Daily Mail
4 Evening Standard
5 Guardian (8 Dec 1972)
6 Herald Tribune
7 Daily Mirror
8 Daily Express
9 The Guardian
10 Daily Mail
11 The Times
12 Evening News
13 The Guardian
14 Evening News
15 Daily Mail
16 The Guardian
17 Daily Mail
18 The Guardian
19 The Guardian
20 Daily Telegraph
21 Daily Mail
22 The Guardian

Chapter 10:
Neil Armstrong

1 'Savour Past Lunar Glories' by John Noble Wilford in the *New York Times*
2 'Moonwalker Recluse' by David Colton in *USA Today*
3 Article by Dick Teresi of *Omni* magazine

General sources

Ten Years Later 'The Moonwalkers' by Lawrence Wright
'Return to Earth' by Buzz Aldrin, Random Books
'Neil Armstrong's Famous First Words' by George Plimpton in *Esquire* magazine

'Savor Past Lunar Glories' by John Noble Wilford, New York Times
'Moonwalker Recluse' by David Colton
'Surfing in Space – people' by Dick Teresi of *Omni* magazine

Buzz Aldrin

1 H. J. P. Arnold, with help from Keith Wilson
2 'Return to Earth' by Buzz Aldrin, Random House

General sources
'Return to Earth' by Buzz Aldrin, Random Books
'Lunar Surface Photography. A Study of Apollo II' by H. J. P. Arnold of *Space Frontiers* published in *Spaceflight*, July 1988
'Who's Who in Space' by Michael Cassutt, G. and K. Hall.

Pete Conrad

General source
'The Men on the Moon come down to Earth' by Daniel Greene, *Observer Magazine*

Alan LaVern Bean

1 'To make the Moon more beautiful' by Patricia Trenner, *Air & Space* magazine
2 'Astronaut Alan Bean' by Susan Adamo, *Starlog* magazine
3 'Astronaut Alan Bean' by Susan Adamo, *Starlog* magazine
4 *Air & Space* magazine
5 'From Astronaut to Artist' article in *Space World*
6 *Chicago Tribune*
7 *Houston Post*

General sources
'Astronaut Alan Bean' by Susan Adamo, *Starlog* magazine
'To Make the Moon More Beautiful' by Patricia Trenner, *Air & Space* magazine
'From Astronaut to Artist' by Bridget Mintz Register, *Space World*
'The Last Word' in *Houston Post*
'Ex-Astronaut wants to leave legacy of art' by Olive Talley, *Chicago Tribune*

Al Shepard

1 Ten Years Later – The Moonwalkers' article by Lawrence Wright

General sources
'The Right Stuff' by Tom Wolfe, Johnathan Cape
'Ten Years Later – The Moonwalkers' by Lawrence Wright

Ed Mitchell

General sources
'The Men on the Moon who came down to Earth' by Daniel Greene, *Observer Magazine*, interview in Spaceline newsletter
'Ex-Astronaut on ESP' – *New York Times*
'Supernatural not Supernatural' by Tom Butler, *Today* newspaper
'Ten Years Later – The Moonwalkers' by Lawrence Wright
Palm Beach Post Times
Article in *Omni* 1984
Letter to author

Dave Scott

1 'Ten Years Later – The Moonwalkers' by Lawrence Wright
2 'Man in Space' – *New Scientist*

General sources
'To Rule the Night', by James Irwin, Hodder and Stoughton
'Ten Years Later – The Moonwalkers' by Lawrence Wright
'A Walk on the Moon' by David Scott
'Man in Space' published by *New Scientiest*

Jim Irwin

1 'To Rule the Night' by James Irwin, Hodder and Stoughton
2 'To Rule the Night' by James Irwin, Hodder and Stoughton
3 'To Rule the Night' by James Irwin, Hodder and Stoughton

General sources
'To Rule the Night' by James Irwin, Hodder and Stoughton
Comprehensive taped interview with author
'Space Explorer's second chance' by Jim Irwin, *Voice*, High Flight Foundation

John Young

General sources
'John Young – One Man's Conquest of Space' by David Shayler, *Spaceflight* magazine
Letter to author

Charlie Duke

1 'Walk on the Moon. Walk with the Son' by Charles Duke
2 'Walk on the Moon. Walk with the Son' by Charles Duke
3 'Walk on the Moon. Walk with the Son' by Charles Duke
4 'Walk on the Moon. Walk with the Son' by Charles Duke
5 'Walk on the Moon. Walk with the Son' by Charles Duke
6 'Walk on the Moon. Walk with the Son' by Charles Duke
7 'The Men on the Moon came down to Earth' by Daniel Greene, *Observer Magazine*
8 'Walk on the Moon. Walk with the Son' by Charles Duke
9 'Walk on the Moon. Walk with the Son' by Charles Duke
10 'Walk on the Moon. Walk with the Son' by Charles Duke

General sources

'Ten Years Later — The Moonwalkers' by Lawrence Wright

'The Men on the Moon came down to Earth' by Daniel Greene, *Observer Magazine*
'Astronaut quits for new challenges' by Nicholas C. Chriss, *Los Angeles Times*
'Walk on the Moon. Walk with the Son' a personal publication provided by Charles Duke

Gene Cernan

General sources

Interview by *Spaceflight News*
'Who's Who in Space' by Michael Cassutt, G. and K. Hall
'The Men on the Moon came down to Earth' by Daniel Greene, *Observer Magazine*
Interview in *Today* newspaper
Article in *TV Guide*, 1985

Harrison Schmitt

1 'Ten Years Later – The Moonwalkers' by Lawrence Wright

General sources

1 'Ten Years Later – The Moonwalkers' by Lawrence Wright
Letter to the author

Index

Note: This Index concentrates on references applying to the space programme, primarily Apollo, and the astronaut-side of astronauts' careers and does not, for example, include references to place names and family associations which are not in the context of spaceflight.